Digital Wellness

Restoring Real-Life Connection
in the Digital Age

Dr. Ali Abdulaziz Al Ali

Table of Contents

Purpose of the Book

To provide a comprehensive, easy to understand guide that enhances digital awareness and helps individuals and families use technology in a healthy, safe, and balanced way, with a focus on psychological, health, social, technical, and regulatory aspects.

Dedication

I dedicate this work to every individual, family, educator, and leader striving to build a healthier relationship between humanity and technology. It is offered in gratitude to those who believe progress must serve human well-being and to future generations whose lives will be shaped by the choices we make today. May it contribute to a more conscious, ethical, and compassionate digital future that empowers people and strengthens societies.

About the Author

Dr. Ali Abdulaziz Al Ali is a specialist in digital transformation and advanced technology. He has extensive leadership experience in developing smart systems and deploying Artificial Intelligence and the Internet of Things in complex work environments, including the energy and services sectors.

"Over the years of my work, I have seen how technology has reshaped daily life, from the way we work, to how we raise children, to our social relationships. Despite the countless benefits of technology, I have also noticed its hidden negative effects on mental health, family cohesion, and quality of life."

From here, my journey with the book "Digital Wellness" began, not only as a theoretical concept, but as an urgent human issue that affects all of us, as parents and educators, as professionals, and as individuals striving to find balance in a world that demands we be *always connected.*

This book is not purely academic, nor is it a conventional awareness message. It is a practical guide grounded in experience, written for anyone who: asks, *"Do I use my phone too much?* "feels their children are drifting away because of devices, or is trying to set digital boundaries without becoming isolated from real life.

Motivations for Writing This Book

My passion for technology, and my deep concern for people, are what led me to write this book. I grew up in the world of technology and witnessed, up close, the leaps of digital development with all the opportunities and challenges they brought. I have seen for myself how technology can make life easier and open new horizons, and at the same time how it can exhaust us and turn us into captives of our screens if we lose control.

Through years of work developing technology products, I felt the growing need to guide usage toward what truly benefits people, in body and in spirit. On a personal level, I went through moments when I realized I was spending excessive time in front of a screen at the expense of my family and my health. I also saw friends and relatives around me experiencing similar digital pressures.

My first motivation is to help people regain control and achieve digital balance in their lives. Whether you are a parent worried about your children's device addiction, a teacher noticing declining student engagement under the dominance of screens, or a young person seeking a way to balance the real and virtual worlds, this book is for you.

In the pages ahead, you will find a practical guide and simple steps, supported by scientific and real-life experience, designed to help you build a healthy relationship with technology without academic complexity or dry technical terminology. I have also made sure the language of the book is close to the reader, and filled with stories and advice that encourage positive change.

In line with the United Arab Emirates declaration of 2025 as the year of the community and 2026 as the year of the family, the importance of this book emerges within an urgent national and humanitarian context. Preserving social cohesion, along with the psychological and social wellness of individuals, has become one of the most pressing priorities of this stage, especially amid the rapid digital changes that have affected the nature of relationships, family bonds, and lifestyles.

Digital wellness is no longer an intellectual luxury. It has become a societal necessity in the face of the challenges of the digital age. This book aims to empower families, educators, young people, and all members of society to build responsible digital awareness that achieves balance between technology and the human being. Through its various chapters, this work offers a knowledge based and community contribution that aligns with the UAE vision of strengthening family cohesion, protecting mental health, and fostering a sense of digital citizenship in a more aware society.

In closing, digital wellness is not an impossible goal. It is an achievable journey when we possess awareness and determination. I invite you to accompany me through the chapters of this book on a journey to explore new ways of managing your digital life and enhancing your health and peace of mind. Let us move forward together toward a balanced life in the age of screens, one in which technology serves us rather than burdens us. The time has come to restore our digital balance and live a happier and more productive life, with balance and awareness.

Introduction

The world around us is no longer what it was two decades ago. Digital transformation has sparked a sweeping revolution that has reshaped our lifestyles, relationships, work, learning, and even our mental health. Today, we are connected around the clock through smartphones, tablets, and computers, making information and communication more accessible than ever before.

Tasks can now be completed quickly, and we can easily communicate with anyone, anywhere, at any time. New opportunities for remote work and distance learning have become widely available. These changes accelerated even further during global events such as the COVID-19 pandemic, which pushed work and education into the online space almost overnight. Yet alongside this remarkable progress, noticeable side effects have begun to emerge. Digital communication has taken over direct human interaction, work time has blended into personal downtime, and screens have made their way to the bedroom and classroom. As a result, many people struggle to focus, experience weaker social connections, and report increased levels of stress and anxiety.

Recent studies show the negative effects on employees' wellness due to the growing difficulty of disconnecting outside working hours. Most participants reported that digital devices interrupt family time, and many spouses noted that smartphones and social media distract them, making it difficult for them to have meaningful conversations with their partners. Despite the tremendous benefits of technology, excessive and unmindful use can harm mental and physical health, weaken family and social bonds, and negatively affect academic and professional performance.

These digital challenges are increasingly visible in everyday life. How often do families gather in the same room, yet each person is silently absorbed in a phone screen? How many teachers stand before their students only to find their attention scattered by a constant stream of notifications? And how many young men and women lose hours of sleep as they scroll endlessly through social media, only to feel anxiety or insomnia when they finally try to rest? Such scenes have become familiar in the age of screens, revealing the real challenges we now face.

A hidden dependence on digital devices has become widespread, with some individuals reaching what is known as "digital addiction"—the excessive and un-reflective use of technology to the extent that other aspects of life are neglected. In this reality, an important question arises: how can we benefit from the advantages of the digital age without becoming victims of its downsides?

To address these challenges and restore balance in our lives, the concept of digital wellness emerges as a thoughtful and practical solution. Digital wellness refers to maintaining overall health in the digital age by adopting positive digital habits that protect psychological, social, and physical wellness, alongside the responsible use of technology. In short, it is the ability to use technology with awareness and moderation, allowing it to serve our lives rather than control them.

This concept includes several interconnected dimensions:

- **The Psychological Dimension:** Supporting mental and emotional health while using technology by managing digital stress and anxiety, maintaining focus, and reducing the negative effects of excessive exposure to information and social media on mood and wellness.

- **The Social Dimension:** Building healthy relationships in the digital age by balancing online interaction with real-world communication. This includes strengthening genuine connections with family and friends, using digital platforms to reinforce rather than weaken bonds, and practicing appropriate digital etiquette.

- **The Physical Dimension:** Protecting physical health within a digital lifestyle by reducing the harm caused by prolonged screen use. This includes maintaining regular physical activity, protecting eyesight from screen strain, and ensuring sufficient, uninterrupted sleep away from electronic devices.

- **The Technical Dimension:** Engaging with technology smartly and intentionally so that we manage it instead of allowing it to manage us. This includes developing digital time-management skills, adjusting device settings to minimize distracting notifications, using online tools that track and regulate screen time, and ensuring privacy and digital security.

- **The Ethical Dimension:** Upholding human values and principles in the digital world just as we do in real life. This involves acting responsibly online, respecting others' privacy, avoiding cyberbully and hate speech, and using digital content in lawful and fair ways. It also includes awareness of technology's broader societal impacts, such as its effects on the environment and social inequality, and behaving in ways that serve the public good.

"Digital Wellness" represents a holistic vision for achieving a healthy alignment between our real lives and the virtual world. It"s more than simply reducing screen-time; it's about building a balanced relationship with technology that supports our values and goals. Through this balance, we can live happier, more mindful lives in the digital age. When we understand these dimensions and work to achieve them, technology becomes a tool that empowers our ambitions rather than a source of pressure.

In the chapters ahead, this book will explore how digital wellness can be achieved step by step through a practical and accessible approach.

Chapter One:

"Modern Technology and its Impact on Daily Life"

In today's world, modern technology has become an inseparable part of our daily routine. Rapid digital transformation has changed the way we live, work, learn, and communicate with others. Smartphones, computers, and tablets are no longer merely entertainment tools; they have become essential means for completing tasks and staying connected.

As reliance on technology continues to grow, we see significant benefits, such as easy access to information and instant communication. At the same time, challenges have begun to emerge, including the difficulty of disconnecting from screens. For example, in a 2019 survey in the United Kingdom, 86 percent of participants believed that technology negatively affects wellness in the workplace due to the inability to switch off and disconnect after working hours. Additionally, 70 percent of people stated that technology sometimes interrupts family time, and 75 percent of spouses reported that smartphones and social media distract them and prevent deep conversations. These figures sound an alarm that achieving digital wellness in our lives has become a necessity.

In this chapter, we will explore how modern technology has affected our daily lives across education, work, health, and social relationships. We will also discuss the evolution of digital technologies such as smartphones, the internet, artificial intelligence, and the Internet of Things, and how people use them in their everyday lives. We will include examples from Gulf countries, especially the United Arab Emirates and Saudi Arabia, on digital transformation, along with models from leading countries in this field, such as Singapore, South Korea, and the Scandinavian countries, for comparison and to draw lessons.

We will examine the positive impacts of these technologies, such as convenience and easy access to information, as well as the negative impacts, such as the risk of digital addiction and social isolation. We will also look at real stories of individuals and institutions that have succeeded, or struggled, in dealing with modern technology. The aim is to understand our new digital reality and pave the way toward digital wellness, meaning achieving a smart balance in the use of technology to protect our health and happiness.

The Evolution of Modern Digital Technology

In recent decades, digital technology has developed in a remarkable way, making our lives today look completely different from what they were one generation ago. The internet has spread from being a limited technology in the 1990s to becoming a global network that connects more than two-thirds of the world's population. Globally, there are now about 5.16 billion internet users, meaning that 64 percent of the world's population is connected online. This massive spread of the internet has made information available to everyone at any time, with the click of a button.

The spread of the internet coincided with a revolution in mobile communication devices. At the beginning of the twenty-first century, mobile phones were traditional and simple. Then, in 2007, smartphones emerged with the release of the first iPhone, bringing together the capabilities of a computer and a phone in one small device.

Today, it is estimated that there are more than 5 billion mobile phone users worldwide, and some advanced countries have reached an almost universal rate of smartphone ownership. For example, in South Korea, the rate of smartphone ownership is about 98 percent among adults, one of the highest rates in the world.

This is no longer limited to phone calls. The smartphone has now become a multifunction device, a platform for social communication, a tool for work and learning, and a means for entertainment and online shopping, among many other uses.

Alongside the internet and smartphones, we have witnessed a surge in other technologies such as artificial intelligence, AI, and the Internet of Things, IoT. Artificial intelligence has moved from research laboratories into commercial and everyday applications. We now find smart systems in everything, from movie recommendation engines on streaming platforms to voice assistants such as Siri and Alexa in our homes, to translation tools and smart photography features on our phones. In recent years, a new type of AI has emerged that can converse and generate content, reflecting a major leap in machines' ability to understand language and interact with humans.

As for the Internet of Things, which refers to connecting devices and objects to the Internet so they become smart, it began with connecting computers and phones, then expanded to include everything: televisions, watches, cars, home appliances, and even light bulbs and thermostats.

Current estimates suggest there are approximately 18 billion internet-connected devices worldwide, from sensors in factories to smart devices in homes, and the number is expected to rise to around 25 to 30 billion devices in the next few years. This means much of what surrounds us has become smart, or at least capable of connection and digital communication.

For example, refrigerators can track their contents and order the missing items themselves. On the other hand, smartwatches monitor our health every second and send data to apps on our phones.

In short, technological development is accelerating at a rapid pace. In just twenty years, we have moved from a world where phones were used only for calls to a world in which we carry a powerful computer in our pockets, the smartphone. We connect to a network that brings together billions of people, and we interact daily with artificial intelligence without even realizing it, through translation, photo editing, and social media, while surrounding ourselves with smart devices at home and at work. This progress has fundamentally reshaped daily life, as we will see in the following sections. Modern technology has become like air; we need it and consume it throughout the day in a wide range of activities. Yet this widespread presence raises an important question: How has technology affected the details of our daily lives in education, work, relationships, and health? Let us explore this, supported by the latest statistics and studies.

Education in the Digital Age: A Revolution in Learning

No sector has undergone as profound a transformation as education over the past decade—all thanks to modern technology. Education today is no longer confined to the walls of traditional classrooms. Digital tools and online platforms have become an essential part of the learning experience.

For example, many schools and universities rely on virtual learning management systems and distance learning platforms to deliver lessons, assign homework, and track student performance. These changes accelerated dramatically during the COVID-19 pandemic, when schools and educational institutions around the world closed, and millions of students were forced to learn from home. At the height of the pandemic, UNESCO data indicated that more than 1.6 billion students in over 190 countries were affected by the disruption of in-person education.

This vast figure, representing about 90 percent of all students worldwide, reflects the scale of the disruption that suddenly struck traditional education, and helps explain why digital learning has become a necessity, not merely an option, in the twenty-first century.

Faced with this reality, educational institutions accelerated their adoption of online learning. Most universities around the world moved their classes online in response to the public health crisis, and statistics from 2023 indicate that around 98 percent of universities worldwide shifted their teaching online following the outbreak of COVID-19.

This rapid transition allowed students to continue their education remotely, thereby ensuring the continuity of learning despite campus closures. At the same time, open online learning platforms such as Coursera and Udemy expanded significantly, creating opportunities for millions of learners to acquire new knowledge through digital means. Today, students can attend lectures delivered by prestigious universities abroad or enroll in specialized training courses without leaving their homes.

This surge in digital education has brought both positives and negatives. On the positive side, technology has expanded learning opportunities in an unprecedented way.

A student in a remote village can now access educational resources and courses from the world's leading universities with the click of a button, contributing to the democratization of education and the spread of knowledge on a broader scale than ever before. Online education also offers a more personalized and flexible learning experience, adapted to the learner's pace and abilities, and it enables self-directed learning at any time that suits them.

Some studies point to broad acceptance of this mode of education. For example, 70 percent of university students believe that online learning can be more beneficial than traditional classroom education, and 95 percent of students recommended online learning to their peers due to the flexibility and value they experienced. These figures confirm the growing satisfaction with digital education among learners, especially among younger, tech-savvy generations.

However, on the other hand, digital transformation has brought new challenges to education. Not all students had the same level of access to the internet or digital devices, which exacerbated educational gaps between social classes and regions. We've seen real cases of families struggling to provide a computer for each child during the period of remote learning, or students forced to rely on a small smartphone to attend long lessons, which had an impact on the quality of their understanding. Teachers also found themselves required to adapt quickly to new learning tools—sometimes without sufficient training, adding further pressure to their workload. And while technology has reduced distances, it cannot fully replace face-to-face classroom interaction. Remote learning may limit direct social contact among students and affect the development of their social skills.

For example, a mother may notice that her child misses interacting with peers and feels isolated, even while attending a "virtual classroom" filled with students on a screen.

Furthermore, the readiness of digital infrastructure varies from one country to another. In advanced countries, the transition to digital education was smoother due to widespread access to high-speed internet in most households. In other countries, distance learning faced obstacles related to weak connectivity or a lack of devices among large segments of the population, creating the challenge of ensuring that no student is left behind.

It is worth noting, however, that the Gulf Arab states invested heavily in building digital infrastructure for education. In the Kingdom of Saudi Arabia, for example, the Ministry of Education launched the Madrasati platform for distance learning during the pandemic. Successfully, it was able to educate around six million students during the 2020-2021 academic year.

Thanks to this national platform, data indicate that about 98 percent of students in the Kingdom used Madrasati during the lockdown period, and the experience received international praise as a successful model of digital transformation in education. In the United Arab Emirates as well, schools quickly adopted a blended learning system and benefited from earlier government initiatives in digital transformation, such as the Smart Learning Project that began before the pandemic. These Gulf examples show that early investment in educational technology paid off during the crisis, reducing disruption to traditional education compared with countries that were not digitally prepared.

In summary, education in the digital age has become a mix of opportunities and challenges. Our daily learning experience has changed fundamentally.

Children now carry a tablet among their school tools rather than books alone, and university students attend academic seminars through virtual platforms instead of lecture halls. These shifts raise an important question: how can we maximize the benefits of technology in learning while avoiding its downsides? Answering this requires awareness and a balance between digital tools and traditional approaches. Here, the concept of digital wellness in education comes to the forefront, and we will return to it later to ensure that digital use supports learning rather than harms it. But first, let us move to another field that has undergone tremendous change due to technology, the workplace.

Work in the Digital Age: Virtual Offices and Blurred Boundaries

Modern technology has transformed the way we work in a way no less revolutionary than its impact on education. Just a few years ago, the traditional routine for most people was to wake up early, commute to the workplace, and spend hours in an office filled with employees. Today, remote work and hybrid work have become a familiar reality for millions of workers around the world. Thanks to widespread high-speed internet and the availability of virtual meeting tools, many people can now complete their tasks and communicate with colleagues from their homes or from any place they choose. This shift did not happen by pure chance. It began before the pandemic as a result of technological progress and changing work culture, but it accelerated dramatically under the global circumstances of the COVID-19 pandemic. When everyone was forced to stay at home, both companies and employees discovered that productivity is possible, and sometimes even better, outside the walls of the traditional office.

Recent statistics reveal the depth of this transformation. For example, in 2023, it was estimated that around one-fifth of the workforce in the United States worked fully from home permanently. In other words, roughly one out of every five American employees does not go to a physical office daily. This percentage represents a major rise compared to the period before 2020, when the share of remote workers in most countries did not exceed a few percent. Overall, the proportion of people working remotely worldwide multiplied several times after the pandemic. More importantly, remote work is no longer merely a temporary measure during an emergency. It has become a preferred career option for many employees. A 2024 survey study indicates that the overwhelming majority of employees want to retain some form of working from home even after the pandemic.

Only 5 percent of employees expressed a desire to return to a full office schedule throughout the week, while 95 percent prefer either fully remote work or a hybrid model. These figures reflect a profound cultural change in how people view work, as they have recognized the high value of the flexibility that technology provides.

On the positive side, digital transformation in the workplace has delivered tangible benefits for both employees and employers. For employees, remote work offers greater flexibility in planning the day. For example, it allows them to save the long hours of daily commuting and invest that time in other tasks or in rest. Many working parents have found that working from home gives them an opportunity to spend more time with their children or manage household responsibilities more effectively, which helps create a better balance between professional and personal life. The absence of a daily commute has also reduced stress and improved physical and mental health, with no traffic congestion and no need to wake up before dawn to make it to work.

Companies have benefited as well. Operating costs have decreased as the need for large office spaces has been reduced, and some surveys have shown that a significant proportion of remote workers felt their mental and physical health improved as a result of this work style.

On the other hand, technology has also introduced challenges into the work environment. Many people today struggle with blurred boundaries between work time and personal time, because the same devices they use to work in the living room are also the devices they use for social communication or relaxation.

A person may find themselves checking their professional email late at night or over the weekend, creating a feeling of being constantly connected to work. This overlap can lead to a form of psychological exhaustion known as "digital burnout," resulting from the lack of separation between professional and private life.

Imagine, for example, a father who finishes a virtual meeting with his team on the computer, only to find that the moment he closes it, his children are asking for help with their online school assignments. As he tries to focus with them, new work-related notifications start pouring in on his smartphone. This kind of constant digital multitasking affects concentration and raises a person's stress level. Studies in workplace settings indicate that frequent interruptions, such as a notification sound or an incoming message, can significantly distract employees. In fact, it may take more than twenty minutes to regain focus fully after each minor digital interruption. Despite the advantages of flexible work, there is a pressing need to learn time management skills and set boundaries for technology use during work in order to avoid exhaustion and loss of productivity.

Another aspect of the digital work experience is the feeling of isolation at times. When an employee works alone away from colleagues, they may miss the informal daily human interaction that used to happen in office corridors or around the coffee table. Some new employees who join a remote role find it difficult to integrate into the company culture or build relationships with the team through a screen alone.

Although there are communication solutions, such as regular virtual meetings for casual social conversation, or occasional face-to-face meetups in hybrid models, human connection through technology remains different from direct interaction.

A real example of this is the story of a teacher who started a new job in a virtual school. After a few weeks, she felt she knew her colleagues' voices through the meeting app, yet she might not recognize them if she happened to meet them in the street. Such paradoxes have become common in the age of virtual offices.

The challenges of the digital age at work are not limited to individuals alone. They also extend to organizational culture and legislation. Many countries have begun to recognize the problem of constant connectivity that digital life imposes on employees, and they have introduced laws that support what is known as "the right to disconnect from work." For example, France passed a law in 2017 requiring large companies to establish a policy regulating managerial communication with employees outside official working hours, to protect work-life balance. In countries such as Belgium, Spain, Italy, and Canada, similar measures have emerged, reflecting global recognition of the need to set healthy boundaries for technology use in the workplace.

In summary, digital technology has fundamentally changed how we earn a living and carry out our daily work. The laptop has become the new office, and the living room has become the meeting room. Many of us reap the benefits of this flexibility, but we also pay a price in the form of new challenges that require awareness and a rethinking of the way we work. Once again, the concept of digital wellness stands out as an important framework for managing these shifts in a way that protects our health and maintains psychological and physical balance.

We will return to this concept later. For now, let us move on to the impact of technology on family and social life.

Family and Social Relationships: Between Digital Closeness and Real-World Distance

Technology has made its way into the heart of family and everyday life, shaping how we communicate with our loved ones and spend time with them. Today, it is rare to find a family without multiple internet-connected devices, smartphones in the hands of parents and children, tablets perhaps for younger kids, and smart TVs streaming content from digital platforms. This new reality has changed the nature of traditional family interaction. In the past, gathering around the table or in the living room was an opportunity to share conversations about the day's events. Now, however, we often see each person absorbed in their own small screen, even when everyone is sitting in the same room.

You may find the father scrolling through work emails, the mother following the latest posts on social media, the son immersed in a video game, and the daughter watching short entertainment clips, all at the same time and in the same place, without exchanging more than a few words.

This phenomenon extends beyond the home environment. Interactions with family and friends have often shifted to mainly digital platforms. Undoubtedly, technology has introduced significant benefits in this context, facilitating connections among family and friends regardless of distance. A son studying in another country can now easily update his mother about his day through a video call, while family chat groups on applications like WhatsApp have emerged as a means to send greetings on special occasions and organize get-togethers. In social situations such as birthdays, some individuals now opt for an electronic greeting card or a voice message instead of a traditional paper card or lengthy phone conversation, which were once commonplace.

These tangible instances illustrate how technology has effectively bridged the gap created by physical distance. Families can now hold virtual gatherings via group video calls, where everyone can see each other, a concept that would have been inconceivable just twenty years ago.

However, digital proximity can sometimes come at a social price. Despite its significance, electronic communication cannot completely replace face-to-face interactions. Over-dependence on texts and online responses may diminish our ability to communicate effectively in person. Some family members might feel overlooked when their loved ones are engrossed in their screens, even when they are sitting right next to them. Numerous parents voice concerns that screen time is taking away from the moments once spent on family conversations or shared activities like group play or outings.

Conversely, children, particularly teenagers, often express that they perceive their parents as being glued to their phones. A recent survey conducted in the United States revealed that nearly half of adolescents believe their parents occasionally become distracted by their devices during conversations with them. This notable statistic indicates that the challenge affects both generations. Everyone, from parents to children, is susceptible to immersing themselves in the digital realm, neglecting their immediate surroundings. In that same survey, approximately 36 percent of parents acknowledged that they also spend more time on their smartphones than they should, suggesting that both sides recognize a disconnect in family communication attributed to technology.

On a brighter note, we now have a greater capacity to share our everyday experiences with loved ones in real-time. Nowadays, it's possible to capture a picture of a child engaging in something amusing and immediately send it to their grandparents online, reinforcing their connection to that moment despite the physical distance.

Technology has also made it easier for individuals to forge new friendships and connections through online platforms. Finding others with similar interests through forums or social media has become a hassle-free venture, leading to the formation of online communities that sometimes evolve into genuine friendships offline. We have all come across stories of individuals who started as online acquaintances and later developed their relationships into face-to-face meetings, marriage, or enduring friendships that extended beyond the virtual space.

Socially, the downside lies in the potential replacement of authentic relationships with what can be perceived as superficial digital interactions. For instance, a young individual might misinterpret the large volume of online interactions, likes, and comments on their posts as indicators of friendship or popularity; however, they may actually lack a true companion who would accompany them on walks or support them during tough times. Prolonged screen time has also limited shared family activities beyond electronic media. It has become commonplace to discover each family member engaging with different content individually on their phones or tablets, rather than gathering together to watch a movie on television as was once customary. Even when family members are physically present, moments are frequently disrupted by quick phone checks or message replies, which disrupt the flow of conversation and diminish focus.

Moreover, excessive exposure to the curated photos and updates of others on social media can lead to unrealistic comparisons. An individual may observe, via their phone, carefully selected images of friends' seemingly perfect and happy lives, and may mistakenly believe their own life lacks excitement or joy, failing to recognize that online shares represent curated highlights rather than a comprehensive view of reality.

In light of these challenges, some parents have started implementing household strategies to mitigate the pervasive influence of technology on relationships. One simple yet common approach is establishing device-free hours each evening, during which everyone sets aside their phones and devices to engage in conversation about their day or play board games together. Such measures signify a genuine effort to achieve a balance between the digital sphere and genuine family life. They serve as a reminder that, despite how much technology virtually unites us, nothing can replicate the warmth of direct human connections.

It is also important to acknowledge that Gulf countries, known for the significant technology usage among their populations, have begun to pay particular attention to this aspect.

In summary, families today benefit from technology in many ways, but it is important to remain mindful of its impact on the quality of our relationships and our ability to be fully present together. Here too, the need becomes clear for awareness of "digital wellness within the family," meaning our ability to use devices with awareness and moderation in a way that strengthens our relationships rather than weakens them. Before moving on to a broader discussion of digital wellness in the **next chapter**, one final topic remains just as important: mental health and focus in an age dominated by screens.

Mental Health and Focus: Wellness Challenges in the Age of Screens

As our reliance on digital devices throughout the day continues to grow, the effects of this lifestyle have become increasingly evident on our mental health and our ability to focus. The constant stream of notifications, frequent switching between on-screen tasks, and prolonged daily exposure to electronic media all affect our brain and psychological state in several ways. Distractibility has become a hallmark of the digital age. A person's mind may jump from an incoming message to a news alert, to a new email notification within minutes, making it difficult to maintain deep focus for long periods. Many people feel that their attention span has shrunk compared to the past, and that sustained reading or completing uninterrupted mental work has become a challenging task when a smartphone is always within reach.

Scientific studies have begun to track these phenomena systematically. Many research findings indicate that excessive screen time is linked to sleep problems and to symptoms of anxiety and depression in some individuals.

When we spend evening hours moving between apps or watching videos before bed, our sleep may be negatively affected by the blue light emitted from screens, which disrupts the body's biological clock. Poor sleep, in turn, affects mood and increases anxiety the next day, creating a vicious cycle of exhaustion.

When it comes to young people in particular, there is growing concern about the effects of technology on mental health. National data in the United States showed that about half of adolescents aged 12 to 17 spend four hours or more per day in front of screens. According to these worrying figures, high levels of screen use are associated with notable mental health concerns. One report found that 27 percent of teenagers who spend four or more hours a day on electronic devices show symptoms of anxiety, while 26 percent exhibit symptoms of depression.

By comparison, anxiety and depression rates were significantly lower among teenagers with lighter use, less than four hours per day. Of course, these numbers do not mean that screens are the only cause, but they do indicate a close association that deserves attention. In everyday life, parents and teachers can notice some of these effects. For example, a student may appear distracted in class because they stayed up late on their phone, or a young person's mood may become noticeably worse if the internet goes down for a short time, as if their life has stopped.

Focus and attention are also victims of the fast digital pace. Every notification that reaches our devices is like a call that pulls our attention away from what we were doing. Most of us have probably experienced this.

You try to complete a task or read a book with concentration, and suddenly you find yourself dropping what you are doing to check why the phone rang or why the screen lit up. The problem is that returning to focus after distraction is not as immediate as we think. A well-known study showed that a person may need about 23 minutes to fully regain concentration after being interrupted. Imagine the amount of time lost and attention scattered if your phone sends an alert every few minutes. Unfortunately, this is the reality for many of us today.

Digital multitasking has become a common habit, such as replying to a message during a meeting or browsing social media while watching a lecture. But research shows that it reduces performance quality and increases mental fatigue. Some people use a vivid term, "monkey mind," to describe the state of constant mental jumping between tasks and thoughts without rest, a state fueled by the accelerated digital rhythm we live in.

Moreover, social media has a clear impact on our mental health. Despite its benefits in staying connected and informed, excessive use can lead to negative psychological patterns such as social comparison and the fear of missing out (FOMO). How many times have you found yourself comparing your life to what you see in the photos of friends and followers and wondering, "Why does their life seem more exciting or happier?" Or felt anxious that you might miss something important because you are not constantly online? These feelings have become common, especially among young people, and have become part of the psychological challenges of the digital age.

Adults are not immune either. An employee may feel pressure to check their email constantly out of fear of missing an important work message, and a mother may feel guilt or inadequacy when she sees the "perfect" images of other mothers on Instagram.

As human beings, we have begun to realize that these platforms have two sides. As awareness grows, calls for moderate and mindful use of social media have also increased.

One powerful real-life example often mentioned in this context is the story of a young man who decided to step away for one day from his smartphone and all social media apps just to try it. He found that he experienced an unusual mental calm and was able to focus on a favorite hobby he had neglected for a long time. Another young man had to install special apps that remind him to stand up and take a break whenever he scrolls for too long without interruption, to protect his physical and mental health.

Such stories have inspired others to try as well, and a wave of "digital fasting" or reducing digital intake has begun to spread as a way to restore mental wellness and reset priorities.

At the same time, it is important to acknowledge the positive side of technology when it comes to mental and cognitive health. Technology is a double-edged sword. Just as it has created challenges, it has also provided innovative solutions. Today, there are apps dedicated to meditation, relaxation, and daily mood tracking, and others that support focus by blocking distractions for set periods, such as apps that temporarily silence notifications or play calming music to aid concentration. Access to mental health support has also become easier online. Remote counseling services and therapy sessions with specialists through video calls have emerged, along with digital support communities that bring together people going through similar experiences to exchange advice and encouragement. In other words, technology can be part of wellness when used wisely.

The real challenge lies in how we balance the significant benefits we gain from technology with protecting our mental health and minds from exhaustion. This leads us directly to the concept of "digital wellness," which addresses this balance, and which we will explore in more detail in the **next chapter.**

Examples from the Gulf and the World in Digital Transformation

To gain a fuller picture of modern technology, it is useful to look at models from different societies around the world and how digital transformation has affected their daily lives, whether positively or through the challenges it brings, and how these experiences compare. Here, we will review examples from the Gulf Arab states, which are among the most technology-engaged societies, and compare them with leading global models such as Singapore, South Korea, and the Scandinavian countries.

The Gulf Area

The Gulf Arab states generally stand out for very high levels of technology adoption and strong digital infrastructure. The United Arab Emirates, for example, is among the world's top countries in internet penetration and smartphone usage. The number of internet users in the UAE reached around ten million in 2023, meaning that 99 percent of the population is connected to the internet. There are also about 9.7 million smartphone users in the UAE, which means that the overwhelming majority of adults have at least one smartphone, and many people own more than one device.

People in the UAE lead a highly active digital life. Statistics indicate that the average time a person spends online each day exceeds seven hours, placing the UAE among the highest countries globally in terms of daily usage intensity.

Gulf governments have actively driven this widespread technological adoption through ambitious digital transformation strategies. For example, the UAE launched smart government initiatives and paperless transactions years ago, and even announced a plan for a paperless government by 2021. Today, citizens and residents in Gulf countries can complete most government transactions through apps and digital platforms, such as renewing official documents or paying bills easily from a phone. In Saudi Arabia, there has also been a strong drive toward digital transformation under Vision 2030. Government e-services have expanded through the Absher platform and others to cover a wide range of sectors, from issuing documents to booking medical appointments.

NEOM City has also been established as a laboratory for future smart cities built on artificial intelligence and the Internet of Things.

In terms of infrastructure, Gulf countries invest heavily in modern technologies. For example, the UAE and Saudi Arabia were among the first countries in the region to launch commercial 5G networks, providing very high internet speeds that supported the use of advanced applications. The UAE also appointed the world's first Minister of State for Artificial Intelligence in 2017, signaling the country's serious commitment to adopting these technologies in developing the economy and government. In Saudi Arabia, the Saudi Data and AI Authority (SDAIA) was established to lead the national strategy in this field.

Socially, people in the Gulf spend long hours on digital platforms for communication and entertainment. In Saudi Arabia, for example, reports have shown that around 49 percent of internet users spend at least seven hours per day online. Smartphone use is dominant, with 99.4 percent of Saudi internet users using their phones to access the internet. Gulf countries also rank among the top globally in per capita usage of platforms such as YouTube, Instagram, and Snapchat. These figures highlight how technology has become a core part of daily life in the Gulf, from education, as we saw with the Madrasati platform in Saudi Arabia, to work, as many companies moved to enable remote work, to shopping, where e commerce has grown significantly in recent years, and even entertainment, such as electronic gaming, where Saudi Arabia has hosted major global tournaments.

Of course, this Gulf leadership in digital transformation also carries challenges similar to those we discussed, such as the need to raise awareness of balanced use and to avoid the negative effects of excessive technology use. We have already seen proactive initiatives in the Gulf to address this, such as launching the National Digital Wellbeing Hotline in the UAE to support families, and enacting laws to protect personal data and safeguard children online in both the UAE and Saudi Arabia. All of this shows that Gulf countries are not only adopting technology, but are also seeking to regulate its pace in a way that achieves "digital wellness" for their societies.

On the other side of the world, there are many countries considered digital leaders, and we can learn from their experiences. Among the most prominent are Singapore, South Korea, and the Nordic countries, such as Sweden, Finland, and Denmark.

Singapore

Singapore is an example of a country that adopted a comprehensive strategy to become a Smart Nation. Its digital infrastructure is among the most advanced in the world. High-speed internet is available almost everywhere, and widespread reliance on e-government services has made completing transactions extremely easy and fast for citizens, such as renewing passports or licenses through a single app called Singpass. Singapore's population has a high level of digital literacy and uses apps in nearly every detail of daily life, from booking taxis to making electronic payments in stores, which became the norm during the COVID-19 pandemic. Singapore also developed the TraceTogether app for contact tracing to limit the spread of the virus digitally, making it one of the earliest global experiments of its kind.

South Korea

South Korea is also a remarkable digital success story. It arguably has one of the highest average internet speeds in the world, and it was among the earliest countries to roll out fiber optic networks from the early 2000s. This enabled an early and widespread internet culture. It is enough to note that 99 percent of adults in South Korea use the internet, and that smartphone ownership is close to 98 percent, as we mentioned, one of the highest rates in the world. Korea is also known for distinctive digital phenomena such as PC bangs, gaming cafés that have been widespread since the 1990s. The government also adopted initiatives such as smart schools and electronic curricula early on. Public culture generally views technology as part of national strength, and the Korean government invests heavily in research on robotics, artificial intelligence, and self-driving vehicles. Yet despite this technological abundance, Korea has not been free of challenges such as internet and gaming addiction among young people, which prompted the government to establish programs and treatment centers to support severe cases of digital addiction. In other words, Koreans also faced the need for balance, despite their strong enthusiasm for technology.

Scandinavian Countries

As for the Scandinavian countries such as Finland, Sweden, Denmark, and Norway, they are consistently ranked among the top countries in indicators of digital quality of life. These countries adopted technology to make citizens' lives easier, while also being known for a culture that values work-life balance. In Finland, for example, digital education is highly advanced. The government provided each student with a computer and an internet plan and integrated coding into school curricula. At the same time, these countries have some of the strongest protections for privacy and digital rights. Denmark and Sweden also developed a comprehensive digital government infrastructure. In Denmark, for instance, you can start a company, pay taxes, or even access your electronic health record through a single portal, and government paper correspondence has been almost entirely phased out.

These countries also use technology to strengthen democracy. For example, Estonia, while not geographically Scandinavian but culturally close, is considered the first country to offer secure nationwide electronic voting in public elections for all its citizens, supported by advanced encryption.

Notably, these Northern European countries achieve a good balance. Their digital infrastructure and services are highly efficient, yet they do not allow technology to take over in a harmful way. Even technology companies in those countries adopt healthy policies for their employees, and many encourage digital detox breaks and flexible working hours that prevent tech-related burnout. This may be one reason they consistently rank at the top of global Digital Quality of Life indicators. According to the 2023 Digital Quality of Life Index report, Finland ranked among the top three countries in the world, and Denmark, Germany, and other European countries also placed in the leading positions.

This suggests they provide high-speed internet at affordable prices, have advanced legislation for privacy protection, and place strong emphasis on digital awareness among citizens.

In summary, when we compare the Gulf with these global models, we find shared priorities, such as investing in technology to develop the economy, education, and services. We also see differences in digital culture. Gulf countries are moving very quickly in adopting and mainstreaming new technologies, and the key challenge for them is to embed a culture of mindful use of this technological abundance, while ensuring that digital divides do not arise within society, for example, between generations or between citizens and expatriates. As for digitally advanced countries like South Korea and Singapore, they began their journey earlier and may have faced certain problems before addressing them over time. Korea, for instance, dealt with the issue of gaming addiction, and Singapore faced the challenge of protecting privacy amid widespread digitalization.

The Scandinavian countries present a compelling model for creating a healthy balance. They use technology to improve efficiency and convenience, yet they have also established rules that ensure our lives do not turn into an exhausting digital race, such as the right to disconnect after work that we mentioned, or community education about the risks of cyberbullying.

These comparisons show that the whole world, despite differences in the speed of digital adoption, shares the search for a point of balance that makes technology a servant of people rather than their master.

Successes and Setbacks in Dealing with Technology

After reviewing different areas of life and how they have been affected by modern technology, it may be useful to pause and consider a few real stories of individuals or institutions that directly experienced the challenges of the digital age, and how they faced them, sometimes successfully and sometimes with difficulty. These real-life examples help bring the ideas to life and shift them from a theoretical framework to tangible situations that could happen to us or to those around us.

A Family Adopts a Screen-Free Day

We begin with the story of a family that decided to address the problem of everyone being absorbed in their phones by introducing a simple rule, one evening per week, without electronic devices. In this family, the father and mother noticed that communication with their teenage children had become superficial and fragmented, as everyone was on a phone or tablet most of the time. So they agreed to dedicate every Friday evening, starting after six o'clock, as completely screen-free family time. Phones would be placed in a box, the TV turned off, and the time spent on shared activities such as cooking together, playing cards, or simply talking.

At first, they faced resistance from the children and perhaps a sense of boredom. But over the weeks, this tradition turned into a cherished time for everyone. One of the children later said that he began looking forward to Friday night because it was the only opportunity during the week when he received his parents' full attention, and they could talk freely without interruptions. The father also said that they discovered many things about their children's school and personal lives during those quiet evenings, things they had previously missed. This story, which was published on an educational blog, spread widely and encouraged other families to adopt the idea, even if only for one day. It shows that simple solutions like this can make a significant difference in restoring family warmth.

An Employee Reclaims Digital Balance

Let us move to the case of someone working in a prestigious position at a technology company who gradually found himself consumed by work around the clock because of constant connectivity. He received dozens of emails every evening after leaving the office and felt immense pressure to respond immediately, even at the expense of time with his family or rest. Work time and personal downtime became completely blended during remote work in the pandemic, and he sometimes worked on his computer until midnight. This person began experiencing symptoms of what we call "digital burnout," insomnia, persistent anxiety, exhaustion, and a loss of enjoyment in the work he once loved.

When the company became aware of the problem of employee burnout, it launched an internal initiative allowing staff to set specific hours during which they would not receive any work messages. In other words, the company adopted a form of "the right to disconnect from email after working hours."

He applied the policy strictly for himself, no email after seven in the evening, and no replies on weekends. At first, he feared how management or clients might react, but he was surprised to find that the company encouraged this approach and did not penalize it. Instead, an emergency duty rotation was assigned to another colleague to handle any urgent matters outside his working hours. Within a few weeks, he began to notice a significant improvement in his mental wellness.

His sleep became regular again, he started setting aside time to exercise in the evening, and most importantly, he regained a normal relationship with his family during the hours that had previously been lost to checking email.

His colleagues say he returned to work in the morning with greater energy and motivation than before. He also became more effective at managing his time during the day because he knew he would not be working at night. This real case, similar to experiences in some European companies that officially adopted restrictions on email after work hours, shows that institutions can play an important role in protecting employees from digital exhaustion, and that everyone benefits in the end.

The UAE Digital Wellness Initiative

We conclude with a real-world institutional and national example from the United Arab Emirates. With the spread of digital addiction patterns and challenges linked to internet use, a pioneering initiative was launched under the name "The Digital Wellbeing Council."

This council aims to strengthen the positives of the digital world and reduce its negatives for society. One of its initiatives was the creation of a free national helpline that allows parents, children, and young people to contact experts and counselors about any problems they face online. For example, if a child experiences cyberbullying on a platform or a parent is concerned about the content their child is viewing, they can call this helpline and receive scientific and practical guidance.

This initiative was described as the first service of its kind in the region to directly support digital wellness for families and the wider community. In its first year, it received hundreds of calls and inquiries, highlighting the scale of societal need. Many callers expressed gratitude for the presence of a trusted entity that offers support on issues that were not easy to talk about previously, or to find solutions for, such as video game addiction among some children, cases of privacy breaches affecting teenagers online, or even simply requesting advice on setting household rules for device use. The success of this experience encouraged discussions about expanding it to other countries.

It is an inspiring example of how technology itself, phones and communication tools, can be used to correct harmful patterns of technology use, in other words, using technology to address the harms of technology, so to speak. It also reflects a leadership vision that recognizes the importance of digital balance and is proactive in adopting policies that achieve it.

Learning from Tech Industry Leaders

It is also worth mentioning a fact that may surprise many people: some pioneers of the technology industry themselves set strict limits on its use for their children. Steve Jobs, the co-founder and former CEO of Apple, for example, did not allow his children to use the iPad when it was launched in 2010, even though it was the tablet device he helped bring to the world and that achieved global success. In an interview, he stated that he limited the amount of technology his children used at home because he believed it was important for them to grow up without complete reliance on screens.

Many other Silicon Valley leaders have done the same. Bill Gates, the founder of Microsoft, did not give his children smartphones before the age of fourteen, and he set limits on their device and gaming time. Evan Spiegel, the co-founder of Snapchat, said that he allows his children only an hour and a half of screen time per week. Even current tech CEOs, such as Sundar Pichai of Google and Tim Cook of Apple, have spoken publicly about the importance of monitoring children's digital use and avoiding excess.

These facts invite us to pause and reflect. If technology experts and the industry's most influential figures are keen to protect their own families from digital overuse, should we not be equally mindful? They serve as very real case studies. Actions speak louder than words. They show that even at the heart of the tech revolution, there is an understanding that moderation is the key.

From these diverse stories, one idea becomes clear: the challenges are digital, but the solutions are human. Technology exists to serve us. When we notice an imbalance, we must make conscious decisions, at the level of the individual, the family, and the institution, to regulate our use. Most importantly, these solutions are possible, not impossible. People like us have succeeded in changing their habits. Companies have adopted more mindful policies.

Governments have launched supportive initiatives. All of this confirms that digital wellness is an achievable goal when efforts come together and awareness rises.

Toward Balanced Use and Life-Giving Technology

After reviewing the key areas of daily life and the impact of modern technology on them, in education, work, family, mental health, and focus, we can clearly sense that digital

transformation has fundamentally reshaped our lives. In the earlier examples, we saw how the day of an ordinary person has become filled with digital interactions, from a lesson received through an online platform, to a work meeting on a virtual app, to a family chat on WhatsApp, to hours spent on social media or in front of a phone screen. There is no doubt that we benefit greatly from technology and appreciate the convenience it has brought us. Yet at the same time, we face real challenges that affect our quality of life, our relationships, and our mental wellness.

As the reader moved through this chapter, they may have felt that some of these changes apply to their own life, or to the lives of people around them. That is expected, because technology has touched everyone in one way or another. The purpose of this discussion was not to criticize technology or glorify it, but to understand our new reality and how to deal with it with awareness. When we realize, for example, that our child is spending more time in front of a screen than before, or that our focus at work has become scattered because of constant notifications, simply becoming aware of that reality is the first step toward positive change.

Here, the concept of "digital wellness" emerges as a central idea that we need to understand and adopt. Digital wellness means achieving a smart balance in our digital lives, using technology in a way that protects our mental and physical health and strengthens our communication and relationships rather than harming them. After seeing how deeply daily life has been affected by digital transformation, it has become essential to develop tools and skills to handle this impact healthily. Just as we learn the principles of healthy living through good nutrition, exercise, and adequate sleep, we now need to learn the principles of digital health. When should we use our devices, and when should we put them aside? How do we build positive digital habits? And how do we protect ourselves and our children from the downsides of the virtual world while still enjoying its benefits?

Digital wellness is not an impossible goal. It is an achievable journey when we have awareness and determination. From real-life examples, we have seen that many people have succeeded in improving their digital balance through simple yet powerful steps. This gives us hope that we can do the same.

In the **next chapter** of this book, we will begin exploring digital wellness in greater detail. We will present practical strategies and recommendations that help us regain control of technology rather than letting it control us. We will look into ways to organize screen time, strengthen focus, and protect mental health amid the overwhelming digital flow. Now that we understand the scale of the impact modern technology has had on our daily lives, it is time to move on to how we can manage this impact to our advantage.

The reader may now feel some surprise or concern about all these digital changes we have witnessed. Yet what is certain is that with proper awareness and guidance, we can turn technology into a tool that helps us live a balanced and happy life, rather than one that makes us lose that balance. This is the core message of our book: your mindful guide to a balanced life in the age of screens.

So, let us move together to the **next chapter**, where we open the gateway to digital wellness and discover, step by step, how to achieve it in practice in our lives. The digital future lies ahead of us, and we must shape it into a healthy and bright future for ourselves and our children. Let us make technology a means that serves us, not a burden on our shoulders. The time has come to restore our digital balance and live a happier and more productive life with balance and awareness.

Let us not become slaves to the digitalization of the world. Let us instead make the most of it.

Chapter Two:

"What is Digital Wellness, and Why Does it Matter to You?"

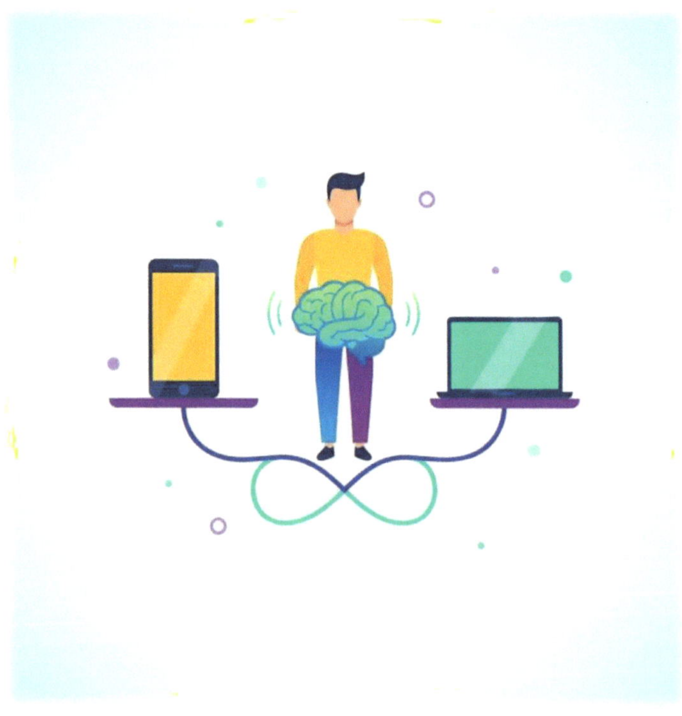

In today's digitally connected world, smart devices and the internet have become an inseparable part of our daily lives. We learn, work, communicate, and entertain ourselves through screens all the time, which makes digital wellness (also known as Digital Wellbeing) an essential concept that everyone should understand. A question may come to mind: "What exactly does digital wellness mean, and why does it matter to you and to you?"

This chapter will take you on a journey to explore the concept of digital wellness, including its various aspects, the distinctions between it and other closely related digital concepts, and the importance of achieving digital balance in our modern lives. We will also review real-life examples from our Gulf communities and from digitally advanced countries, along with success stories of individuals and institutions that adopted digital wellness practices. We will discuss common everyday situations that reveal the impact of the absence of this balance. We will offer you some simple exercises and activities to help you assess your digital awareness and strengthen your digital wellness.

Digital wellness is everyone's focus. It is a matter of concern for parents because it helps them raise a conscious generation that can guarantee a positive and safe utilization of technology. It matters to teachers because it enables them to guide their students to benefit from digital age tools in learning without falling into distraction or exhaustion. It also matters to young people because it protects their health and balance amid daily electronic life and helps them achieve their ambitions without screens standing in their way.

A Comprehensive Definition of Digital Wellness

Digital wellness is a state of healthy balance in a person's digital life. It focuses on the overall impact of technology on an individual's psychological, social, and physical health, including their digital practices. Digital wellness can be defined as: "the impact that digital technologies and devices have on the user's mental, physical, emotional, and social health." This definition suggests that digital wellness is not only about "how much time we spend in front of screens," but also about the "quality of that use" and how it affects us in every aspect. It is a concept that is centered on building and maintaining balance between digital life and real life effectively. In other words, digital wellness is having a healthy and positive relationship with technology. It means we clearly recognize both the benefits and the drawbacks of digital engagement, and we can manage our use of technology in a way that protects our health and happiness.

Psychological and Social Aspects

Digital wellness includes the psychological aspect, which relates to our mental health and emotions during and after using technology. For example, does browsing social media make you feel anxious or lead you to compare yourself with others? Do video games make you feel stressed or frustrated? These are questions tied to your mental health in the digital world. Digital wellness, along with the social aspect, means how technology affects your relationships with others and your social participation.

Excessive device use may lead to isolation and weaker communication with family and friends, while balanced use can strengthen connections, such as staying in touch with loved ones across distances.

Physical and Technical Aspects

Digital wellness encompasses not only psychological and social dimensions but also physical health considerations. Prolonged periods of sedentary behavior while engaging with computers or smartphones can adversely affect the body, leading to issues such as eye strain, cervical and lumbar discomfort, and sleep disturbances attributed to excessive exposure to blue light, particularly at night. A notable finding from a recent study indicated that 28 percent of participants reported a detrimental impact on their sleep quality when accessing digital content prior to bedtime.

The technical aspect of digital wellness pertains to an understanding of technology and the employment of tools and settings designed to mitigate adverse effects. For instance, utilizing application time limits or configuring smartphones to "Do Not Disturb" mode during nighttime hours are straightforward technical strategies that can significantly enhance digital wellness.

Digital wellness represents a holistic concept that encompasses mental health, emotional stability, social connections, physical wellness, and the technical proficiency required to navigate one's digital landscape. This framework serves as a tool for individuals to assess their quality of life in the digital era, aimed at achieving a balanced existence between the virtual environment and reality. Digital wellness doesn't limit itself to psychological and social aspects. It also extends to physical health. Sitting for long hours in front of a computer or phone affects the body in many ways, including eye strain, neck and back pain, and sleep disturbances due to prolonged exposure to blue light at night. A real example is that 28 percent of participants in a recent study felt that viewing any content on their devices before bed negatively affected the quality of their sleep. The technical aspect of digital wellness relates to understanding technology itself and using tools and settings that help reduce negative effects. For example, learning how to use app time limits or setting your phone to "Do Not Disturb" at night are simple technical practices that enhance your digital wellness.

In summary, digital wellness is a comprehensive concept that includes your mental health, your emotions and peace of mind, your social health, your relationships and sense of belonging, your physical health, your body and sleep, and your technical skills in managing your digital life. It is a framework that helps you evaluate the quality of your life in the digital age and improve it by achieving a healthy balance between the virtual world and the real world.

The Difference Between Digital Wellness and Other Digital Concepts

Some people may confuse digital wellness with other common digital concepts, such as cybersecurity, digital privacy, or even general mental health. However, it is important to understand the differences clearly. Each concept has its own focus, even though there are overlaps.

- **Digital Wellness vs Cybersecurity:**

Cybersecurity is concerned with protecting computer systems, networks, and data from attacks and digital breaches. It is about information security and protecting privacy from external threats such as viruses or hacking. Digital wellness, on the other hand, focuses on your own safety and health while using technology. In other words, you may have a secure system protected from hacking and strong cybersecurity, yet still suffer from exhaustion or stress due to excessive device use, meaning low digital wellness. Cybersecurity is an important element of safe digital life, but it does not cover the psychological, social, and physical aspects of our daily technology use. For example, online harassment or personal data exposure is a digital security risk, while digital addiction, distraction, and weak concentration are digital wellness issues. In short, the first protects you from external online risks, while the second protects you from internal overuse and the negative impact of technology.

- **Digital Wellness vs Digital Privacy:**

Digital privacy refers to protecting individuals' personal data and sensitive information while using technology and the internet. In other words, it focuses on your right to control your information, what is shared, and with whom. Digital wellness is broader. It concerns how technology use affects you as a whole. For example, you may be very careful about privacy, you do not share personal information, and you use strong passwords, yet at the same time, you spend long hours online, which affects your health or relationships.

Digital privacy is part of the overall picture, and it is also important for feeling safe and comfortable in the digital world, but digital wellness includes additional aspects such as your psychological state during use, managing screen time, and building positive digital habits. One example is that oversharing details of your life on social media may put your privacy at risk, and it may also cause stress and anxiety about your self-image in front of others. The first is a digital privacy issue, the second is a digital wellness issue.

- **Digital Wellness vs General Mental Health:**

Mental health, or psychological health, is a broad concept that includes a person's emotional state, their ability to cope with life's pressures, and their capacity to realize their potential. Digital wellness is related to mental health, but it is more specific in the context of technology. In other words, it focuses on your mental wellness while you use digital technologies. For example, if you experience anxiety or depression in general, this falls under traditional mental health. But if that anxiety is partly driven by your use of social media, such as feeling isolated or constantly comparing yourself to others, then digital wellness becomes relevant as it addresses the issue within a digital context.

Digital wellness also includes elements that go beyond the psychological aspect to include the social and physical aspects, as mentioned earlier. Therefore, digital wellness can be seen as an extension of mental health in the digital age.

The objective is to ensure that technology enhances our wellness rather than undermining it. For instance, the World Health Organization asserts that mental health is an integral component of human wellness. Digital wellness endeavors to promote this state within the digital realm by utilizing technology in ways that support mental health rather than detract from it. It is crucial to acknowledge that prioritizing mental health, such as by seeking professional assistance when necessary or engaging in relaxation techniques like mindfulness, can be incorporated into an individual's digital wellness strategy if technology appears to have a negative psychological impact.

In summary, digital wellness pertains to the quality of one's digital experiences and overall health while engaging with technology. Cybersecurity focuses on the safety and integrity of data and systems, while digital privacy is concerned with safeguarding personal information and preserving the right to confidentiality online. Mental health encompasses a broader range of issues relating to psychological wellness, independent of its association with technology. Recognizing these distinctions enables a comprehensive approach to address each area effectively and prevent any aspect from being overlooked. Achieving digital wellness necessitates strong cybersecurity measures to protect our digital environments, respect for privacy, and a commitment to overall mental health.

The Importance of Digital Wellness in Modern Life

The question, "Why does digital wellness matter to me?" can be answered by examining the reality of contemporary life. In the age of the digital revolution, individuals spend a significant portion of each day connected to the internet or interacting with digital devices. As a result, digital wellness has become an essential component of overall quality of life. The following real-world examples highlight the importance of digital wellness at the level of individuals, communities, and nations, both within the Gulf region and across the globe.

The Gulf Arab States: The UAE and Saudi Arabia as Examples

The Gulf states were among the early adopters in recognizing the importance of digital balance. For instance, the United Arab Emirates established a dedicated Digital Quality of Life Council in 2020 to promote digital wellness among its citizens. This national council aims to develop policies and programs that ensure a healthy balance between digital life and real life, spread the values of responsible digital behavior, and raise digital awareness across all segments of society.

In addition, the UAE launched a digital wellness helpline, the free number 80091, to provide advice and guidance to parents and individuals on everyday challenges in the digital world. These initiatives reflect the seriousness with which the government addresses issues such as internet addiction and cyberbullying before they escalate, as well as its commitment to digital quality of life as an integral dimension of modern human wellness.

In the Kingdom of Saudi Arabia, a national awareness program titled "Sync for Digital Wellbeing" was launched in 2021 under the supervision of the King Abdulaziz Center for World Culture (Ithra). The program aims to promote healthy digital practices and reduce the negative effects of excessive technology use on individuals and society. What distinguishes Sync is that it is grounded in specialized research conducted in collaboration with universities and research centers.

These studies revealed important findings. For example, 50 percent of Generation Z participants reported experiencing fatigue, lack of sleep, and headaches as a result of excessive technology use. In addition, 41 percent of participants indicated difficulty carrying out daily tasks due to reduced focus caused by constant digital distractions. These figures sounded a local and global alarm and confirmed the urgent need for intervention to promote a healthier culture of screen use.

The Sync initiative has implemented wide-ranging awareness activities, including an annual international summit that brings together experts to discuss the latest research on digital balance, technology addiction, and related topics. As a result, Saudi Arabia has emerged as one of the region's leading models in advancing digital wellness.

Digitally Advanced Countries: Finland, South Korea, and Canada

The importance of digital wellness is not limited to the Gulf region. It has also become a key priority on the agendas of technologically advanced countries worldwide, although the motivations and contexts may differ.

Finland

Finland is widely recognized as a global leader in education. In recent years, it has also acknowledged the importance of maintaining balance in the use of technology within schools. After a period of widespread adoption of digital learning tools, Finnish educational research revealed unexpected outcomes: as the use of digital devices in classrooms increased, students' learning outcomes declined.

A researcher at the University of Helsinki identified Finland's heavy reliance on tablets and computers as one of the factors contributing to the country's decline in international PISA rankings. The issue was not the weakness of technology itself, but rather that excessive use made it easier for students to become distracted. As a result, several Finnish schools began reassessing their technology policies by limiting device use in classrooms and restoring balance between digital and traditional learning methods.

Even in a technologically advanced country like Finland, there is growing recognition that digital overuse can negatively affect young people's focus and learning capacity. The Finnish experience demonstrates that digital wellness is essential for maintaining educational quality and outcomes, and that technology should remain a means to support learning rather than an end in itself.

South Korea

South Korea is among the world's most advanced countries in terms of internet infrastructure and the widespread use of smart devices. It is common to see even primary school students carrying smartphones. However, this rapid progress has been accompanied by serious challenges, including high rates of internet and electronic gaming addiction among young people.

The Korean government was among the first to recognize the severity of this issue and to respond through a national digital wellness framework. In 2011, it introduced legislation commonly known as the "Cinderella Law," which prohibits individuals under the age of sixteen from accessing online games after midnight. Although some teenagers found ways to bypass the law, it sent a clear message that the wellness of young people takes precedence over the profits of the gaming industry.

In addition, the government established treatment and rehabilitation centers to support individuals suffering from severe digital addiction. These include nature-based camps designed specifically for adolescents showing signs of internet addiction. Thousands of students participate annually in programs lasting three to four weeks, which involve physical, social, and outdoor activities away from screens. The aim is to rediscover enjoyment in real life and develop coping skills that do not rely on the internet.

The South Korean experience illustrates how the absence of digital wellness can escalate into a national crisis requiring unconventional solutions to protect an entire generation from becoming lost in the virtual world. Encouragingly, these efforts have begun to yield positive results as public awareness increases and discussions around healthy technology use become more widespread.

Canada

Canada holds a leading global position in promoting digital wellness. In the Digital Wellbeing Index (2024), which evaluates countries on how effectively they maximize the benefits of technology while reducing its negative impacts, Canada ranked first among 35 assessed countries. It achieved the highest overall score across digital wellness indicators.

Canada outperformed countries such as the United Kingdom and Australia in areas including digital physical wellness, particularly in managing screen time and maintaining physical activity. This achievement can be attributed to Canada's comprehensive approach, which combines community awareness with evidence-based guidelines for healthy technology use.

For example, Canadian health and education authorities provide clear recommendations for parents regarding appropriate screen time for children and adolescents. These guidelines emphasize setting firm limits, such as restricting digital entertainment for teenagers to no more than two hours per day, to protect mental and physical health. Canadian schools also integrate lessons on safe and healthy digital behavior into their curricula.

This institutional awareness is reflected in international survey results such as the "Digital Wellbeing Index," in which the United Arab Emirates ranked 16th globally, and Saudi Arabia ranked 27th. These rankings indicate growing awareness in both countries, while also highlighting opportunities for further development compared to leading nations. Overall, Canada's experience demonstrates that sustained investment in digital education and community-wide digital health initiatives produces tangible outcomes, resulting in a more balanced and productive digital society.

These examples from the Gulf region and around the world confirm a fundamental truth: digital wellness has become an urgent necessity rather than a luxury. It is a critical enabling factor for successful digital societies. Countries that prioritize digital wellness tend to cultivate communities that are healthier, more productive, and more socially connected, whereas neglecting it can lead to health, educational, and social crises. While recognizing the importance of digital wellness at the governmental and institutional level is an essential first step, its real impact depends on how it is translated into daily practices within individual lives and families—a topic explored in the next section.

Case Studies: Real Success Stories in Achieving Digital Balance

To move beyond theories and statistics, it is useful to examine real-life stories of individuals, families, and institutions that have successfully adopted digital wellness principles and achieved measurable positive outcomes. These case studies illustrate the meaningful change that becomes possible when people choose to regain control of their digital lives.

A Week Without Screens: A Family Reconnects

In one inspiring example, a mother invited her family to participate in a week-long challenge to completely disconnect from digital devices in order to reassess their technology habits. Surprisingly, the children, aged six and eight, experienced little discomfort during the absence of screens. On the contrary, they became noticeably calmer, and their overall mood improved.

The mother recounts that she initially expected her children to feel bored or struggle to adapt. Instead, they spent time playing outdoors with the family dog, jumping on the trampoline, and engaging in physical and imaginative activities rather than video games. Interestingly, the change extended to the parents as well. After turning off the television and putting away their phones in the evenings, the parents found more time to talk with each other and with their children, and they even returned to reading books—an activity they had not done in a long time.

The mother describes the experience by saying, "Without phones or Netflix, we watched the sunset and talked for a long time. I felt as if time slowed down and returned to us." By the end of the week, the family realized that much of the time they believed they lacked had actually been consumed by screens with limited real benefit. This case illustrates how even a short digital detox can strengthen family bonds and improve emotional wellness for both children and adults.

A Community-Based School Project in Dubai: The 21-Day Challenge

As part of institutional efforts to promote digital wellness, a pioneering initiative emerged in Dubai led by the EdRuption Foundation in collaboration with schools and families under the name "The Digital Bridge." This participatory research project included workshops for parents and students, followed by a 21-day challenge during which participants committed to healthy digital habits. These habits included setting daily device-free times at home and avoiding phone use for at least one hour before sleep.

The outcomes were remarkable. After three weeks, parents reported improved sleep quality, increased screen-free family time, and stronger family relationships. Participating students reported a 100 percent sense of pride in their improved ability to manage screen time. Measurements showed that parents' sleep quality improved by up to 37 percent due to reduced device use before bedtime.

Perhaps most significantly, many families reported experiencing their longest and deepest conversations in years after removing digital distractions. One father reflected after attending a "Vision and Values" workshop with his son, "We have not spoken this honestly in years. I feel like I am seeing my child again." Several schools later used the findings to develop longer-term policies on student phone use, involving students in decision-making rather than imposing unilateral bans. This case demonstrates how collaboration between schools and families can generate rapid and lasting positive change when digital wellness is treated as a shared responsibility.

The Right to Disconnect: An Institutional Policy for Employee Wellness

In professional environments, some institutions and companies have begun to recognize that employee productivity and wellness are closely linked to maintaining healthy digital boundaries. One prominent example is France's "right to disconnect" law, introduced in 2017, which grants employees the right to refrain from responding to work messages outside official working hours. Following this model, some European companies implemented measures such as disabling email servers at night to prevent after-hours communication.

These initiatives led to reduced burnout and increased job satisfaction without negatively affecting productivity. In fact, when employees felt that their personal time was respected, their focus and performance during working hours improved. This example highlights that digital wellness is not limited to children or youth but is equally critical in adult professional life.

Digitally exhausted individuals cannot perform effectively or creatively, whether in academic or workplace settings. As a result, forward-thinking institutions now promote cultures that respect digital boundaries by discouraging after-hours messaging, encouraging email-free breaks or vacations, and offering training in digital time management. These policies serve as large-scale success stories, demonstrating that investing in digital wellness leads to healthier, more focused, and more productive workforces.

Across these diverse examples, from families and schools to corporations and governments, a common lesson emerges: achieving digital wellness is both possible and rewarding. While circumstances and strategies may vary, meaningful change occurs when individuals and communities recognize the value of balance, supported by collective will and thoughtful action. These success stories offer practical inspiration for every reader to begin taking small but impactful steps toward improving their own digital wellness, as explored next.

Everyday Situations That Show the Effects of Missing Digital Wellness

Unfortunately, many of us experience some of the negative effects of digital imbalance in our lives without directly linking them to technology use. Below are some common scenarios that show how a lack of digital wellness can harm us, supported by recent statistics.

Digital Burnout

Have you ever reached the end of a workday feeling that your brain is completely drained simply from sitting in front of a screen all day?

Digital burnout is a state of physical and psychological fatigue that results from prolonged and excessive use of digital devices without sufficient breaks. In the United States, for example, about 67 percent of employees said they sometimes experienced digital burnout, and 22 percent reported that this feeling had become frequent for them. Not only that, but more than 82 percent of remote workers recently said they had experienced signs of technology-related exhaustion, such as fatigue, headaches, and difficulty concentrating.

Digital burnout does not affect employees alone. Students, homemakers, and many others are also vulnerable in our time. Common symptoms include persistent tiredness, reduced motivation for work or study, head and eye pain, and sleep disturbances due to constant screen exposure. Health organizations have warned that if this condition continues, it may lead to deeper health problems over time, such as depression or high blood pressure. Digital burnout is like a battery that keeps losing power without being recharged. So if you find yourself feeling exhausted despite little physical effort, review your digital habits immediately and look for balance.

Distraction and Reduced Focus: Do you find it hard to finish a task or read a single page without your hand automatically reaching for your phone to check notifications? If the answer is yes, you are not alone. Constant notifications and switching between multiple tasks on digital devices can significantly scatter attention. Studies suggest that a person needs, on average, 15 to 20 minutes to fully regain focus after being interrupted by a notification or message. Imagine the impact when we receive dozens of notifications in a single hour. In a recent Saudi study, 41 percent of participants said they had begun to struggle with completing their daily tasks because of reduced focus caused by constant digital distractions.

On a broader scale, a survey in the United States found that the average person checks their phone about 58 times a day. Some reports raise this number to more than 100 times per day among young people, meaning they check their phones roughly every five to ten minutes. This behavior keeps our minds in a constant state of anticipation for updates, which reduces our ability to concentrate deeply, immerse ourselves in tasks, or even enjoy the present moment. So if you feel that your memory is not what it used to be, or you struggle to meet deadlines, the cause may be repeated digital distraction, which can be managed through strategies such as turning off non-essential notifications or setting specific times to check your phone instead of keeping it in front of you all day.

Weaker Social Relationships

One of the most unfortunate scenes of our time is when people gather in the same physical place, yet each person is absorbed in their own private digital world through a phone. This common phenomenon is known as phubbing, meaning ignoring the person in front of you by being preoccupied with your phone. Scientific studies have described phubbing as a small-scale form of social exclusion, where the neglected person feels isolated and rejected. In a survey of 143 adults in the United States, 46 percent reported experiencing phubbing from their life partners, and 23 percent said that phones caused tangible problems in the relationship. It is therefore not surprising that phubbing is linked to lower satisfaction in marital or family relationships and to increased feelings of loneliness and jealousy.

Even having a phone on the table during a conversation with a friend, without actually using it, was enough in one experiment to make participants feel that intimacy and warmth were lower in that dialogue. In other words, simply anticipating that we might pick up the device sends a signal to the other person that our attention is not fully with them.

On the other hand, a British study revealed that when we have our phones with us, we spend 42 percent less time interacting with those around us, because our attention is divided between reality and the screen.

Even more strikingly, 53 percent of people worldwide admitted that they would rather have their smartphone than enjoy the company of a close friend. This shocking statistic reflects how deeply phones have come to fill the emotional space of relationships for some people. These figures and realities invite us to pause and reflect: do we give our loved ones enough attention when they are present? Digital wellness, in its social dimension, means restoring humanity to our communication. Eye contact, a smile, and genuine listening matter far more than any virtual like or fleeting message. Relationship weakening caused by devices may not appear immediately, but it accumulates over time, leaving us with a sense of distance from the people closest to us.

Academic Decline and Poor Study Performance

It is no secret that students today face major challenges in maintaining focus amid constant digital temptations.

When a student does not enjoy digital wellness, the smartphone can shift from a learning tool into a fierce competitor to books and studying. Recent studies link excessive screen time to lower academic performance. A scientific review covering multiple studies found that adolescents who spend more than seven hours a day in front of screens are 40 percent less likely to achieve high grades compared to those who spend less than two hours. The reasons are many, including late-night device use leading to insufficient sleep and weaker comprehension, and distraction during lessons, preventing information from being properly absorbed. We also saw the Finnish example earlier, where excessive use of educational technologies without clear boundaries contributed to a decline in student outcomes despite the strength of the education system.

In addition, many students struggle with the habit of digital multitasking while studying, such as browsing social media while reviewing lessons. This has been shown to reduce efficiency and extend the time needed to complete assignments. Digital exhaustion can also leave a student physically present in the classroom while mentally drained and unable to focus or participate.

The aim here is not to place all the blame for academic decline on technology, but it is clear that our digital habits strongly influence learning outcomes. The good news is that improving a student's digital wellness, such as setting specific times for digital entertainment, keeping the phone away during study, and getting enough device-free sleep, can quickly improve focus and results. Modern education has begun to recognize this. Some schools in France and parts of Canada, for example, have banned smartphones in classrooms entirely to improve the learning environment. It is a reminder that, however useful technology may be, it must remain in the service of education rather than obstructing it.

These everyday situations are not imaginary. Most of us have experienced them to varying degrees. The good news is that we can correct course if we identify the problem and acknowledge it. The first step toward addressing any imbalance is awareness. Now that we have reviewed these effects, it is natural to ask: How do I know my level of digital awareness? And how can I regain balance if I find a problem? This is what we will help you with through some simple exercises next.

Exercises and Activities to Assess and Strengthen Your Digital Awareness

Moving from theory to practical application is the key to creating real change in our digital habits. Here are some simple exercises and activities you can do to evaluate your digital awareness and enhance your digital wellness.

Treat them as a self-assessment or a personal challenge, and answer honestly with yourself:

- **Calculate Your Digital Time:**

Start by observing yourself. How many hours do you spend each day on your phone or computer outside work or study? You can use time tracking apps or the Screen Time feature on your phone to find out. Is it more than you expected? If the number is higher than you thought, this is a sign that you may need to regulate your daily digital intake.

- **The Short Disconnection Test:**

Try putting your phone on airplane mode for one hour during the day, or leave it in another room away from you. How do you feel during that hour? Do you quickly feel anxious or bored? Or did you enjoy focusing on another activity without distraction? Your level of comfort or discomfort in this simple experiment reflects the degree of your reliance on your phone and your constant need for digital connection. It is not a problem to acknowledge that now. Many of us are in the same boat, and awareness is the first step toward change.

Quick Evaluation Questions:

Answer **"Yes"** or **"No"** to the following,

1. Do you take your phone to bed and use it right before sleep or as soon as you wake up?

2. Have you ever felt upset or anxious because you could not check your notifications during a meeting or class?

3. Does your internet use affect your ability to complete tasks, such as delaying study or work because of prolonged browsing?

4. Have family members or friends ever complained that you are "always busy on your phone" when you are with them?

5. Have you ever tried to reduce your use of a social media platform or a particular game and found it very difficult to do so?

If your answer is **"Yes"** to several of these questions, this is a sign that you may need to create a plan to improve your digital balance.

- **A Technology Free Day (Digital Detox Day):**

Choose a day, or even half a day, during the coming week and make it a period free from personal technology use. Use your phone only for essential needs, such as an urgent call, and completely avoid browsing the internet or using social media platforms. On that day, write down notes about your activities and feelings. You may finally find time for a long-delayed hobby, go out for a walk, or have a long conversation with a family member. Many people who try this exercise discover that they have far more time and mental calm when digital distractions are removed. Repeat this exercise regularly, weekly or monthly, as a way to reset yourself.

- **Set Your Rules and Commit to Them:**

Sit alone and create three digital rules or boundaries you want to follow for your wellness. For example, "I will keep my phone away during family meals," or "I will stop scrolling after 11 p.m.," or "I will not use social media during study or work hours." Write these rules on a piece of paper and place it somewhere visible as a form of commitment. You can also involve a family member or a close friend to hold you accountable. Start by applying one new rule every week or two until it becomes a habit, then add more over time.

You will be surprised how these simple steps can give you back a strong sense of control and make technology feel less dominant over your day.

These activities are primarily meant to raise your self-awareness and help you diagnose your digital habits. No one knows your situation better than you do, so be honest and patient with yourself. If you find it difficult, remember that you do not need perfection. Simply trying and improving gradually is already a win. And if you succeed in a step, reward yourself by choosing an enjoyable alternative to screens, such as going out with a friend, exercising, or reading. You deserve to be the master of your devices, not their prisoner, and these exercises are your tools to achieve that.

The Core Value of Digital Wellness and How It Connects to the Rest of Your Life

At the end of this chapter, we return to the opening question: What is digital wellness, and why does it matter to you? We have seen that digital wellness is not merely a theoretical idea. It is a balanced lifestyle in the age of technology. It is the ability to benefit from the vast advantages of the digital world while avoiding the traps of overuse, addiction, and distraction. It means giving our minds, bodies, and inner lives their rightful share of rest, focus, and genuine human relationships, even though devices are everywhere around us.

The importance of digital wellness lies in the fact that it touches everything. It affects your mental health and peace of mind. It affects your productivity and creativity in study or work. It affects the quality of your relationships with the people you love. It even affects how you see yourself and how satisfied you feel with your life. It matters because you are a human being who deserves a better life, a life in which technology is a tool for building and support, not a force that undermines your health and happiness.

In connection with the rest of the book, digital wellness forms the foundation on which other topics are built. The upcoming chapters, whether they address digital parenting for children, online safety and privacy strategies, or the development of future digital skills, all rest on the principle of digital wellness, which helps us use these skills and knowledge with wisdom and balance. For example, there is little value in teaching a child to code apps if they are addicted to gaming in a way that harms their health. The latest smartphone iteration provides no substantial advantages for students who struggle with balancing entertainment and academic commitments. Digital wellness can be conceptualized as a navigational tool that assists individuals in addressing challenges encountered in daily digital interactions. Considerations arise, such as the utility of specific applications versus their potential to become a distraction, whether to accept employment that demands continuous digital connectivity, or how to foster familial happiness amidst an incessant influx of screens. In these scenarios, the principles of digital wellness, namely awareness, moderation, and responsibility, serve as a framework to facilitate informed decision-making.

Ultimately, it is imperative to recognize that the journey towards digital wellness is a continuous personal endeavor. There may be instances of faltering, where one finds themselves engaged in prolonged, aimless scrolling; however, such occurrences are acceptable. The critical factor is the ability to regain awareness and recalibrate one's course. Establishing realistic objectives and acknowledging progress, regardless of its magnitude, is essential. Over time, individuals will find themselves exhibiting greater control and satisfaction in their digital interactions. A sense of accomplishment is derived from concluding a day having achieved planned tasks without succumbing to digital distractions. Moreover, dedicating meaningful time away from devices fosters the development of significant personal relationships. These aspects represent the invaluable rewards associated with digital wellness.

Digital wellness is of paramount importance as it pertains fundamentally to individual wellness in the contemporary digital landscape. By prioritizing such wellness, one invests not only in their personal growth but also in the welfare of their loved ones. Subsequent chapters will explore how parents can implement the principles of digital wellness to cultivate a balanced digital environment for their children, alongside an examination of how these principles enhance personal cybersecurity through conscientious digital practices, among other specialized subjects. The concept of digital wellness will be shown to underpin all topics discussed, establishing it as a foundational element for advancing a healthy and secure digital future for ourselves and forthcoming generations.

Chapter Three:
"Digital Addiction and Time Management"

In an age in which technology has become intertwined with our daily lives in unprecedented ways, it has become increasingly easy to immerse ourselves in the digital world to the point of excess. Many people now spend long hours each day using smartphones, social media platforms, and the internet, raising serious concerns about digital addiction and the ability to manage time in a healthy and balanced manner.

This chapter seeks to address these concerns directly. It presents a clear and precise definition of digital addiction and explains how it differs from simple overuse. It also examines the key signs of digital addiction, along with its psychological, physical, and behavioral symptoms, in order to help readers, recognize when technology use crosses a critical threshold.

In addition, this chapter explains—clearly and accessibly—how digital addiction affects the brain and behavior. Drawing on recent statistics from the Gulf region and around the world, it identifies the platforms and applications associated with the highest rates of excessive use. At the same time, the discussion does not overlook the human dimension. Real-life examples illustrate how digital addiction affects children, adolescents, and adults from academic, social, and psychological perspectives.

The chapter then moves from understanding to action by presenting practical and evidence-based strategies for managing time and reducing overuse. These include techniques such as the 20/20/20 rule, the Pomodoro Technique, and structured approaches to building digital balance. It also highlights leading awareness initiatives and national campaigns in the United Arab Emirates and Saudi Arabia that aim to address digital addiction and encourage healthier relationships with technology. Finally, it shares real experiences and success stories of individuals and families who managed to break the cycle of digital addiction and significantly improve their quality of life.

All of these elements lead to a central conclusion: achieving digital wellness requires self-awareness, discipline, and intentional choices in how we engage with technology.

Defining Digital Addiction and the Difference Between Overuse and Addiction

Digital addiction refers to a state of compulsive attachment to digital technologies, such as the internet, smart devices, or electronic gaming, where usage becomes uncontrollable and begins to negatively affect multiple areas of a person's life. Some experts describe it as compulsive or problematic internet use that leads to significant impairment in a person's ability to function across different life domains over an extended period. In simple terms, the individual loses control over how much time they spend on devices or digital platforms, even when they are aware of the harm it causes.

This condition differs fundamentally from simple overuse. Not every case of prolonged screen time qualifies as a clinical addiction. Long hours alone do not indicate addiction unless certain defining elements are present, which are discussed in the following sections.

It is therefore important to distinguish between excessive technology use and clinical digital addiction. A person may spend many hours in front of screens due to work requirements, academic responsibilities, or even entertainment. This constitutes overuse and may have negative effects, but it does not become an addiction unless disruptive symptoms emerge. Clinical digital addiction—sometimes referred to as internet addiction disorder—is characterized by loss of control over use, accompanied by addiction-like symptoms such as distress when access is restricted, increasing tolerance, and the neglect of academic, professional, or social responsibilities because of technology.

Although excessive internet use has not yet been formally recognized as an independent disorder by the World Health Organization or the American Psychiatric Association, modern medical classifications have acknowledged specific conditions, such as gaming disorder, within the International Classification of Diseases (ICD-11). This distinction reinforces the idea that overuse alone is insufficient for an addiction diagnosis. There must be a clear functional impairment in a person's daily life resulting from their digital behavior.

In summary, overuse involves spending long periods on digital devices, which may lead to fatigue, reduced focus, or weaker social relationships, yet remains partially under the individual's control. Digital addiction, by contrast, is a condition in which the individual loses control entirely. They continue to use technology despite knowing its harmful consequences and feel compelled to return to screens repeatedly. At this stage, digital addiction becomes a genuine mental health concern that requires awareness and intervention.

It is also important to note that the medical and psychological communities continue to debate the precise definition and diagnostic criteria for digital addiction. Experts have not yet reached a universal consensus, and discussions continue about whether it should be classified as a separate disorder or as a symptom of underlying psychological conditions. What experts do agree on, however, is that problematic technology use becomes a serious issue when it overwhelms essential life responsibilities—such as work, education, and social relationships—and when repeated attempts to reduce use fail despite negative consequences.

Signs and Symptoms of Digital Addiction

The signs of digital addiction can generally be categorized into psychological, physical, and behavioral dimensions. Together, these indicators serve as warning signals that technology use has moved beyond normal engagement and has become a harmful dependency.

Behavioral and Psychological Signs

One of the most prominent indicators is the inability to control or stop technology use. The addicted individual repeatedly returns to their device without a clear purpose and experiences a strong compulsive urge to check their phone or remain online. They may lie about the amount of time spent using digital platforms or attempt to conceal their behavior from others.

Neglecting responsibilities and previously enjoyed activities is another serious warning sign. Individuals may ignore academic or professional duties, withdraw from social interactions, and abandon hobbies in favor of constant digital engagement. Over time, academic performance or work productivity often declines as attention becomes increasingly consumed by screens.

Tolerance is another defining feature, meaning the individual must spend more time online or engage with increasingly stimulating content to achieve the same level of satisfaction. Emotionally, attempts to reduce or stop device use may trigger anxiety, irritability, or distress. When access to the internet is interrupted or devices are removed, reactions may resemble withdrawal symptoms seen in substance addiction. Many individuals also use the digital world as an escape from real-life stress, finding temporary comfort online while becoming less capable of coping with offline challenges.

Physical Signs

Excessive screen use without adequate breaks often results in noticeable physical symptoms. Eye strain is among the most common complaints and may include dryness, redness, or blurred vision. Digital addiction also disrupts sleep patterns, as many individuals stay up late using devices and experience persistent daytime fatigue.

Headaches, neck pain, and back pain frequently develop due to poor posture and prolonged sitting. A sedentary lifestyle associated with excessive screen use may contribute to weight gain, reduced physical fitness, and lower energy levels. Some research also links digital addiction to weakened immune functioning, largely due to sleep deprivation and insufficient physical activity.

Additional physical symptoms include carpal tunnel syndrome caused by extended keyboard or mouse use, as well as pain in the thumbs and hands from constant smartphone typing. While these symptoms may initially appear mild, they often worsen over time if unhealthy digital habits persist.

Behavioral Changes in Daily Life

Digital addiction often manifests in behaviors that are easily observed by others. Social withdrawal is common, with individuals preferring online interaction over face-to-face communication. They may avoid social gatherings altogether or attend physically while remaining mentally absorbed in their devices.

Mood fluctuations are also frequently linked to device use. Individuals may appear energized or content while online but become irritable, anxious, or emotionally flat when disconnected. Some develop emotional numbness or lose interest in activities that do not offer the rapid stimulation provided by digital platforms.

Another key sign is a distorted sense of time. Individuals may lose track of how long they have been online and feel shocked when they realize hours have passed unnoticed. Daily routines may be disrupted, including delayed meals, skipped self-care activities, or neglected family responsibilities. Adolescents with severe overuse may forget to eat or drink during extended gaming or scrolling sessions.

In advanced cases, individuals may violate personal values or rules to gain additional screen time, such as using devices secretly at night or accessing the internet during work hours. Taken together, these indicators help parents, educators, and individuals distinguish between a strong interest in technology and a level of dependence that requires intervention.

It is important to emphasize that the number of hours spent online alone does not define addiction. The decisive factor is the impact of digital use on a person's health, behavior, and life balance. Someone may work eight hours a day on a computer without being addicted if they can disconnect afterward and maintain a healthy routine. Another individual may spend fewer hours online, yet experience serious disruption to sleep, studies, and relationships. Thus, assessment focuses on loss of control and functional impact, not merely screen time.

How Digital Addiction Affects the Brain and Behavior

From a scientific perspective, digital addiction shares many similarities with traditional forms of addiction, such as substance or gambling addiction. Engaging in pleasurable digital activities—scrolling social media, watching videos, or playing games—triggers the release of dopamine in the brain's reward centers. Dopamine is the neurotransmitter responsible for feelings of pleasure, motivation, and reward.

Human brains evolved to release dopamine in response to survival-enhancing behaviors such as eating, social bonding, and learning. Modern technology, however, can exploit this system. Many applications and games generate frequent and intense dopamine bursts, closely resembling the effects of addictive substances. Notifications, likes, comments, and in-game rewards each deliver small but powerful dopamine hits that create immediate gratification. With repeated exposure, the brain becomes accustomed to this heightened stimulation.

Over time, excessive dopamine release can disrupt the brain's reward system. To restore balance, the brain reduces its sensitivity to pleasure, causing the individual to feel less satisfaction from the same digital activities. This explains why addicted users often increase usage time or seek more stimulating content. This process, known as tolerance, fuels a vicious cycle of escalating use and diminishing reward.

Brain imaging studies suggest that excessive internet or gaming use may reduce activity in the prefrontal cortex, the region responsible for impulse control, planning, and decision-making. As a result, individuals struggle to resist urges and regulate their behavior.

Digital activities can also produce sharp emotional fluctuations. Pleasure during use may be followed by sudden drops in mood once the activity stops, a phenomenon sometimes described as a temporary dopamine deficit. This can leave individuals feeling bored, restless, or emotionally low, prompting them to return quickly to their devices to restore a sense of comfort.

Importantly, digital platforms are intentionally designed to reinforce this cycle. Features such as infinite scrolling, vivid notification colors, and reward-based game mechanics are engineered to capture attention and sustain engagement. The brain's natural attraction to novelty and surprise ensures continuous dopamine release, keeping users trapped in repetitive loops of consumption. One expert famously described smartphones as delivering "digital dopamine" directly and continuously to the brain's pleasure centers.

These neurological changes inevitably affect behavior. Everyday activities such as reading, studying, or spending time with family may feel dull compared to screen-based stimulation. In some cases, individuals develop anhedonia—the reduced ability to experience pleasure in ordinary life.

Excessive digital use has also been linked to increased anxiety and depression. Social comparison on social media, exposure to idealized lifestyles, sleep disruption, lack of exercise, and reduced human connection all contribute to emotional imbalance. Concentration and memory are also affected, as constant notifications and multitasking train the brain toward distraction. Tasks requiring sustained attention become more difficult and mentally exhausting.

In summary, digital addiction alters brain chemistry in ways that reinforce dependence and weaken self-control. Individuals become more impulsive, less patient, and increasingly driven by immediate gratification, while their mental wellness deteriorates.

Understanding these mechanisms empowers individuals to take corrective steps. Specialists note that extended breaks from highly stimulating platforms—sometimes referred to as dopamine fasting—can help reset the brain's reward system and restore enjoyment in everyday activities. Although such breaks require patience and discipline, their long-term benefits for brain health and overall wellness can be substantial.

The Platforms and Applications Most Associated with Excessive Use: Statistics from the Gulf and the World

In today's technology-driven world, not all digital platforms are equally capable of capturing—and holding—users' attention for long periods. Some applications are deliberately designed in ways that make disengagement difficult, increasing their addictive potential. Recent statistics clearly show that social media platforms, electronic games, and video streaming services account for the largest share of users' daily screen time worldwide.

To put this into perspective, data from 2023 indicate that the global average daily time spent on social media is approximately two hours and twenty-four minutes per person. While this figure represents a global mean, usage varies widely across countries. Some nations exceed this average by a significant margin. For example, users in the Philippines—traditionally among the highest worldwide—spent more than nine hours per day online in 2022. Although this figure declined slightly to around nine hours and fourteen minutes more recently, it remains strikingly high and illustrates how deeply digital life dominates daily routines in certain societies.

Turning to the Gulf region, the pattern is similarly concerning. Statistics consistently show that internet and social media use in Gulf countries ranks among the highest globally. In the United Arab Emirates, for instance, the average daily time spent online is approximately seven hours and twenty-nine minutes, placing the country well above the global average of six hours and thirty-seven minutes among working-age users. Social media use alone accounts for about three hours per day, according to 2022 reports.

The situation in Saudi Arabia closely mirrors this trend. Saudis spend an average of three hours and six minutes per day on social media, a figure notably higher than the global average. Additional data from 2022 indicate that 79 percent of the population actively uses social media, with an average daily use of three hours and sixteen minutes. Together, these figures confirm how deeply digital social interaction has become embedded in Gulf societies.

When examining which platforms consume the largest share of users' time, some regional differences emerge, but the overall pattern remains consistent worldwide. TikTok currently leads globally due to its highly engaging short-video format, followed by YouTube, which dominates longer-form video consumption. Other major platforms—including Facebook, Instagram, X (formerly Twitter), and Snapchat—continue to attract large and loyal user bases.

In the Gulf region, messaging and video-based applications occupy a particularly central role. In the UAE, WhatsApp is the most widely used social application, largely because it serves as an essential communication tool. It is followed by Instagram and Facebook in terms of activity and user engagement. In Saudi Arabia, distinct preferences are evident. Snapchat enjoys exceptional popularity, especially among young people, due to its immediacy and privacy features, which align closely with local cultural norms. TikTok is also highly popular among younger age groups, while X (Twitter) remains influential as a platform for real-time news, debate, and public discussion.

Beyond social media, electronic gaming represents another major digital space associated with excessive use. Games such as Fortnite, PUBG, and Minecraft attract millions of young users who often spend long hours immersed in virtual environments. As discussed earlier, the World Health Organization has officially recognized gaming disorder as a medical condition, based on documented cases in which individuals spend most of their day gaming while neglecting academic, social, or professional responsibilities. In the Gulf, online gaming is particularly widespread due to high-speed internet access and near-universal smartphone ownership among youth. Stories frequently emerge of students neglecting their studies as a result of gaming obsession, highlighting the seriousness of the issue.

It is also essential to consider device accessibility, as addiction potential is strongly linked to availability. In Gulf countries, smartphone penetration and internet connectivity are nearly universal. In the UAE, approximately 99 percent of the population has internet access, and social media account penetration exceeds 105 percent, reflecting the fact that many individuals maintain multiple accounts. Similarly, in Saudi Arabia, 99 percent of the population uses the internet, with the vast majority also active on social networks. Such high levels of digital access create ideal conditions for excessive use: devices are always within reach, and digital services are woven into nearly every aspect of daily life, from education and work to entertainment and shopping.

From a global perspective, the past decade has witnessed a steady increase in screen time. Although there was a modest decline in 2022, when average daily internet use dropped to around six hours and thirty-seven minutes, possibly due to a partial return to pre-pandemic routines, this figure remains historically high. Overall, global internet use has increased by more than half an hour per day over the past ten years. This long-term upward trend clearly signals that the digital world is consuming an ever-greater share of our time, intensifying the need for awareness, effective time management, and balance.

In summary, the platforms most strongly associated with excessive use share common design features: endless content feeds, continuous recommendations, and interactive group-based engagement. Social media thrives on constant updates and notifications, video platforms automatically queue one clip after another, and games rely on continuous rewards and challenges. In Gulf countries—where a large youth population coincides with strong enthusiasm for technology—usage rates rank among the highest globally. While countries such as the Philippines and Brazil often top global rankings with nine to ten hours of daily internet use, nations like the UAE and Saudi Arabia follow closely behind, with averages exceeding seven to eight hours per day, including several hours on social media alone.

These figures raise an unavoidable question: are we truly using these platforms for our benefit, or are they increasingly using us?

The Impact of Digital Addiction on Children, Adolescents, and Adults

Digital addiction does not discriminate by age. Its effects extend across childhood, adolescence, and adulthood, manifesting differently at each stage of life. The following sections present realistic examples and everyday scenarios to illustrate how excessive digital use affects individuals academically, socially, and psychologically.

Children: Innocence Put to the Digital Test

For young children, the digital world is a double-edged sword. Tablets and smartphones provide entertainment and educational content, yet excessive exposure can disrupt healthy development.

Consider a seven-year-old who consistently refuses outdoor play or interaction with peers, choosing a screen instead. Over time, such behavior can weaken attention span, reduce classroom focus, and undermine academic readiness.

Children accustomed to instant digital gratification often struggle with traditional learning environments that require patience and sustained attention. Creativity may also suffer. Instead of inventing games or imagining stories, children consume ready-made digital content, limiting opportunities for imaginative play.

Social consequences quickly follow. A child who prefers screens to social interaction may withdraw from peers, avoiding group play during family gatherings or school activities. This withdrawal hampers the development of essential social skills such as sharing, negotiation, and conflict resolution. Psychologically, mood instability becomes common. Many parents report that children appear calm and absorbed while using devices but become irritable, aggressive, or emotionally distressed when screens are removed—responses that closely resemble withdrawal symptoms.

One mother shared a revealing experience. She allowed her children, aged five and eight, unlimited screen time on weekends so she could rest. Over time, removing devices became nearly impossible without emotional outbursts. When she eventually decided to eliminate screens temporarily, the first days were difficult. However, the outcome surprised her. The children began inventing games, drawing, building puzzles, and playing cooperatively. Arguments decreased, and the home environment became calmer. As she reflected, "I watched my children rediscover their imagination. They became more creative, cooperative, and emotionally balanced."

This example highlights an important truth: boredom without screens often becomes the gateway to creativity, not a threat to wellness.

Teachers increasingly echo these concerns. Many report that young students are becoming more distracted and less responsive in classrooms, a trend often linked to habitual exposure to fast-paced digital content. Language delays have also raised alarms among pediatricians, who emphasize that active conversation—not passive screen viewing—is critical for early language development. In this sense, digital addiction steals valuable developmental time that cannot easily be recovered.

Adolescents: Constantly Connected, Yet Increasingly Vulnerable

Adolescents are perhaps the most vulnerable group. Teenagers naturally seek social connection and identity formation, and social media has become the primary arena for both. While these platforms provide visibility and interaction, they also introduce significant risks.

Academically, excessive digital use often leads to declining performance. A typical scenario involves a student who spends hours gaming or watching videos late into the night, arrives at school exhausted, and struggles to concentrate. Assignments are postponed, grades decline, and frustration builds. Time management becomes increasingly difficult as the phone takes priority over responsibilities.

Socially, many digitally addicted teenagers lead double lives: vibrant online identities alongside passive real-world engagement. Family relationships may suffer as teenagers isolate themselves in their rooms, interacting more with screens than with parents. Attempts to regulate usage often result in conflict, resistance, and emotional volatility. Psychologists note that forced restrictions can trigger aggression or depressive symptoms, particularly when emotional dependence on technology has developed.

Psychological risks are equally serious. Excessive social media use has been linked to higher rates of anxiety and depression, especially among teenage girls. Social comparison, fear of missing out (FOMO), and cyberbullying contribute to emotional distress. In some cases, digital addiction serves as an escape from deeper psychological struggles.

One case study described a teenage boy who spent most of his day gaming. Initially labeled as "addicted," therapy revealed underlying ADHD and depression. After treating these conditions, his gaming time decreased significantly, his academic performance improved, and he reintegrated socially. This case illustrates a crucial point: digital addiction in adolescents often masks deeper emotional needs.

Adults: Productivity at Risk, Relationships Under Strain

Contrary to common belief, adults are not immune to digital addiction. Many struggle silently with compulsive phone use, often at the expense of productivity and personal relationships.

Professionally, excessive screen use reduces focus and efficiency. Employees who frequently scroll during work hours fall behind on tasks and may face consequences. Some organizations have responded by blocking social media sites or enforcing phone-free meetings. Similarly, university students may postpone assignments while losing hours to digital distractions, ultimately affecting academic outcomes.

Within families, digital addiction can erode emotional connection. It is now common to see families sitting together while each member is absorbed in a personal screen. Over time, meaningful conversation declines, leading to emotional distance between spouses and between parents and children. In more severe cases, constant digital engagement has contributed to marital conflict and even divorce.

Psychologically, digitally addicted adults often experience chronic stress and burnout. Constant connectivity creates pressure to respond, stay updated, and remain available. Many report anxiety when separated from their phones, sleep disturbances, and emotional exhaustion. Physically, prolonged screen use contributes to neck and back pain, eye strain, weight gain, and reduced overall wellness.

One woman in her late twenties described reaching burnout after averaging eight to nine hours of daily phone use. A six-day digital detox retreat became a turning point. Upon returning, she set strict boundaries—no phone after 8 p.m., no phone near the bed, and screen-free mornings. As a result, her screen time dropped significantly, sleep improved, focus returned, and she rediscovered forgotten hobbies. Her experience demonstrates that recovery and balance are always possible.

Effective Strategies to Manage Time and Reduce Excessive Use

After understanding the nature of digital addiction and its wide-ranging effects, the next and most important question naturally arises: how can we regain control of our time and our lives in a world dominated by screens? Fortunately, managing digital overuse does not require extreme measures. Experts and everyday users alike have identified simple, realistic strategies that can be gradually integrated into daily routines. The following approaches are among the most effective and widely recommended.

The 20/20/20 Rule: Eye and Mental Relief

One of the simplest yet most effective strategies is the 20/20/20 rule, commonly recommended by eye specialists to reduce digital eye strain. The principle is straightforward: for every 20 minutes of screen use, take a 20-second break and look at something approximately 20 feet (6 meters) away. In practice, this can be easily implemented by setting a gentle reminder while working or studying.

When the reminder sounds, pause your task, lift your eyes from the screen, and focus on a distant object, such as a window or the far wall of the room. This brief interruption allows the eye muscles to relax, helping prevent headaches, dry eyes, and visual fatigue. Equally important, it gives the mind a short pause, restoring attention and reducing mental overload. Many individuals who adopt this habit report improved focus and reduced exhaustion. Beyond eye health, the rule serves another purpose: it disrupts continuous attachment to screens and encourages conscious breaks. Whenever you forget, remember the simple formula—20-20-20.

The Pomodoro Technique: Structured Focus and Productivity

The Pomodoro Technique offers an effective solution for those who struggle with procrastination or mental fatigue. Developed by an Italian student in the 1980s, the method divides work into short, focused intervals—typically 25 minutes, known as a "Pomodoro"— followed by a 5-minute break. After completing four Pomodoro's, a longer break of 15 to 30 minutes is taken.

This structure works because 25 minutes is long enough to make progress yet short enough to maintain mental clarity. Knowing that a break is approaching increases motivation and reduces the urge to check the phone mid-task. Over time, this approach trains the brain to focus deeply for short periods, making large tasks feel more manageable. Pomodoro is especially effective for studying, office work, and creative tasks, as it encourages regular breaks while preventing burnout. Whether using a dedicated app or a simple timer, many people are surprised by how much they accomplish in a few focused cycles compared to long, unstructured screen time.

Setting Schedules and Creating Device-Free Time Blocks

Reclaiming control over digital habits requires intentionally protecting certain parts of the day. Establishing device-free time blocks is a powerful step. Experts strongly recommend avoiding screens at least one hour before sleep and one hour after waking up. Evening screen use interferes with sleep quality, while starting the day without immediate digital stimulation allows the mind to wake gradually and calmly.

Meals and family time should also become technology-free zones. Placing phones aside during lunch or dinner encourages conversation, strengthens relationships, and restores meaningful presence. Study and homework periods should be protected in the same way. Parents can support this by encouraging children to leave devices in another room during study sessions.

Another effective practice is setting fixed times for social media use. Instead of checking platforms repeatedly throughout the day, designate two specific periods—perhaps in the afternoon and evening—for limited use. This approach reduces constant distraction while maintaining a sense of connection.

Turning Off Notifications and Reducing Digital Triggers

Notifications are among the strongest drivers of compulsive phone use. Managing them effectively can dramatically reduce screen time. Begin by disabling notifications for non-essential apps. Social media, games, and news applications rarely require immediate attention. Keeping alerts only for important calls or messages makes the phone quieter and far less tempting.

Organizing the phone's home screen can also make a difference. Move distracting apps off the main screen and place them in folders or on secondary pages. Replace them with practical or calming apps, such as health tools, reading apps, or spiritual resources.

Another surprisingly effective technique is switching the phone to grayscale mode, which removes color from the display. Bright colors stimulate the brain, while black-and-white screens reduce visual reward, making prolonged scrolling less appealing—especially at night.

Using Digital Time Management Tools

Ironically, technology itself can help address excessive use. Most smartphones now include built-in features such as Screen Time or

Digital wellness, allowing users to monitor usage and set limits. By establishing daily caps for specific apps, users become more aware of their habits and are encouraged to stop once limits are reached.

Features such as Downtime, Sleep Mode, or Focus Mode further support self-control by restricting access to distracting apps during specific periods. For example, scheduling downtime from 10 p.m. to 7 a.m. protects sleep routines and prevents late-night scrolling. Focus Mode is especially useful during work or study, as it temporarily silences interruptions until tasks are completed.

Motivational tools such as Forest also offer creative solutions. By linking focus time to the growth of a virtual tree, users are rewarded for staying off their phones, turning self-discipline into an engaging challenge.

Personal "No Phone" Rules and Behavioral Boundaries

Setting clear personal rules strengthens digital discipline. Examples include: no phone at the dining table, no phone in bed, no phone during family visits, and no phone during the first hour of work or study.

Another useful habit is the "break-the-loop" rule. If you notice yourself scrolling aimlessly for more than ten minutes, close the app and engage in a physical action—stretching, walking, or drinking water. Keeping devices physically out of reach during work or study further reduces temptation and improves concentration.

Healthy Breaks and Non-Digital Alternatives

Healthy screen habits also depend on how breaks are used. After every 20 to 30 minutes of focused screen work, take a short break that does not involve another device. Stand up, stretch, walk briefly, or speak with someone nearby. After longer periods, step away completely for ten minutes to reset both mind and body.

Equally important is replacing screen time with meaningful offline activities. Reading, exercising, cooking, drawing, volunteering, or spending time in nature provide healthier forms of satisfaction. When enjoyable alternatives exist, reducing screen dependence becomes far easier.

Gradual Reduction and Sustainable Balance

The goal of digital wellness is not total disconnection, but balance. Abrupt change often fails. Gradual reduction is far more effective. Reducing screen time by one hour per week allows habits to adjust naturally. Tracking progress, reviewing weekly screen-time reports, and rewarding success help maintain motivation.

Many individuals report significant improvements through small changes, such as charging the phone outside the bedroom, starting mornings phone-free, or using airplane mode during focused work. These practices return control to the individual rather than the device.

Ultimately, consistency and flexibility matter most. Not every strategy suits everyone. The key is to experiment, commit, and adapt. When digital habits align with personal values, technology becomes a tool that supports life—rather than one that consumes it.

Initiatives and Awareness Campaigns in the Gulf to Address Digital Addiction

Across the Gulf region, governments and institutions have increasingly recognized the importance of promoting digital balance and reducing the harmful effects of digital addiction on individuals and families. As a result, a wide range of awareness initiatives, national programs, and community-based efforts have been launched to address this growing challenge.

In the following section, we highlight prominent examples from the United Arab Emirates and the Kingdom of Saudi Arabia, two countries that have taken notable steps toward building healthier relationships between people and technology.

Leadership in Digital Quality of Life

In the United Arab Emirates, concern for digital wellness is not marginal; it is recognized at the highest levels of leadership. The country has clearly acknowledged the need to protect society in the digital space and to ensure that rapid technological advancement does not come at the expense of human wellness. One of the most prominent examples of this commitment is the establishment of the Digital Quality of Life Council.

This council plays a central role in coordinating national efforts to enhance digital quality of life. Its work includes reviewing relevant legislation, proposing policies, and launching initiatives and programs that promote positive, safe, and balanced technology use across society. Through these efforts, the council seeks to embed digital wellness as a shared cultural value rather than a personal responsibility alone.

Among the council's recent achievements is its collaboration with major technology companies to introduce protection features for younger users. In June 2025, the council announced the launch of a new safety feature in partnership with Meta, Facebook, and Instagram, designed specifically to protect teenagers on Instagram. This initiative, the first of its kind in the region, uses advanced algorithms to detect harmful or inappropriate content and prevent it from appearing in teenagers' feeds. It also includes parental monitoring tools that strengthen privacy while supporting children's psychological safety. This initiative reflects the UAE's seriousness in working directly with global technology companies to create a safer and more balanced digital environment for future generations.

Beyond platform-level protections, the UAE has also integrated digital wellness into its broader national agenda for happiness and quality of life. A national policy for digital quality of life has been launched to support the development of a safe, positive, and human-centered digital society. As part of this effort, the Digital Quality of Life Support Line was announced in July 2025.

This service operates as a dedicated helpline staffed by specialists who provide advice, guidance, and practical support to parents and community members facing digital challenges. Through a specialized call center, the service helps families address issues such as children's gaming addiction, excessive screen use, and exposure to cyberbullying. By offering direct consultation rather than abstract guidance, this initiative acknowledges that families need real, accessible support to manage the complexities of digital life.

At the community level, a range of local campaigns has also emerged. Initiatives such as the Child Digital Safety campaign aim to promote safe internet practices among children and parents alike. Schools, parenting centers, and educational institutions regularly organize workshops that address the risks of excessive device use and teach practical strategies for online safety. Nonprofit organizations and private-sector partners also play an active role. For example, Snapchat partnered with the Digital Quality of Life Council to launch the Charter for Child Digital Flourishing, through which technology companies pledged to provide safer digital environments for children. These partnerships illustrate a collaborative model of digital governance that brings together government, industry, and civil society.

Media-based awareness initiatives further strengthen these efforts. One notable example is the national campaign Our Digital Safety, launched in 2021 by the Dubai Foundation for Women and Children to raise awareness of cyberbullying, digital addiction, and related online risks. The campaign provided guidance materials for parents and children and promoted healthy digital behavior. In another initiative, the Telecommunications and Digital Government Regulatory Authority (TDRA) organized educational workshops at the GITEX exhibition, in cooperation with TikTok, focusing on family online safety, data privacy, and parental control tools.

Overall, the UAE's approach reflects a clear belief that digital transformation must progress alongside social values and public awareness. As Sheikh Saif bin Zayed stated, "Digital wellness goes beyond infrastructure and smart services; it focuses on protecting generations and promoting responsible engagement with technology." This vision emphasizes capacity building, policy development, practical support, and broad community engagement as the foundations of positive digital behavior.

Toward a Balanced Digital Society

In Saudi Arabia, government institutions have also taken steady and deliberate steps to raise awareness of digital risks and promote responsible technology use, in alignment with Vision 2030. Central to these efforts is the Ministry of Communications and Information Technology (MCIT), which has launched campaigns targeting children, parents, and the wider community. These campaigns explain both the benefits and risks of the digital world and offer guidance on safe and effective technology use through workshops, publications, and simplified family-oriented resources.

Educational institutions have played a similarly active role. Schools and universities regularly organize awareness weeks that address issues such as smartphone addiction and excessive screen use. For instance, one Saudi university held a health awareness week focused on the risks of smartphone addiction, featuring student-led lectures, digital booklets, and peer-generated guidance materials. Initiatives like these empower young people to engage critically with their own digital habits rather than simply receiving instructions from authority figures.

From a mental health perspective, the National Center for the Promotion of Mental Health has published awareness materials explaining digital addiction, its symptoms, and ways to address it. Through its website and social media platforms, the center has shared simplified infographics and educational content, signaling continued attention to the issue. The Ministry of Health has also contributed by providing parental guidance on children's screen use, including strategies for managing screen time and recognizing early signs of addiction. These efforts reflect growing recognition that digital addiction is not merely a behavioral issue, but a public health concern.

Nonprofit organizations and national initiatives further complement government efforts. The "Digital Giving" initiative, for example, focuses on strengthening digital culture and awareness, while also addressing practical issues such as screen time management. Security agencies have contributed as well. The General Directorate of Narcotics Control has included warnings about so-called "digital drugs" within its awareness campaigns, linking emerging online risks to broader discussions of addictive behavior in cyberspace.

At the community level, family counseling centers have begun to address digital addiction directly. The Family Development Association in Al Ahsa reported treating more than 50 cases of digital addiction through family counseling programs. This reflects growing awareness within civil society and highlights the role of family-based interventions in restoring balance.

Recreational and youth-focused initiatives also play an important role. Campaigns such as A World Without Addiction, launched by Princess Nourah bint Abdulrahman University, and "A Week Without a Screen" competitions organized by youth centers encourage young people to rediscover offline activities through sports, arts, and social interaction.

Saudi Arabia has also integrated digital education into school curricula, addressing safe and balanced technology use, digital citizenship, and digital health. These efforts aim to raise a generation that understands both its digital rights and responsibilities.

Overall, Saudi initiatives focus on prevention, awareness, and empowerment. While national strategies provide direction, universities, schools, families, and associations contribute actively to confronting digital addiction before it escalates.

Chapter Four:
"The Psychological and Social Impact of Technology"

Psychological Effects of Excessive Technology Use

In today's digital age, prolonged exposure to technology—such as smartphones, computers, and social media platforms—has been increasingly linked to a range of negative effects on mental health. When technology use becomes excessive and unregulated, it can disrupt an individual's psychological and emotional balance. Numerous cases have reported symptoms of anxiety, depression, sleep disturbances, and low self-esteem among people who spend long hours in front of screens. Below is a simplified overview of the most prominent psychological effects associated with excessive technology use.

Persistent Anxiety and Stress

The constant flow of notifications, messages, and negative news through digital applications can place individuals in a state of continuous psychological pressure. Ongoing exposure to disturbing or alarming news content may intensify feelings of stress and anxiety, particularly among individuals who already experience elevated baseline anxiety levels. In addition, the fear of missing out (FOMO) plays a significant role. Many users feel compelled to check their phones repeatedly, remaining in a state of constant alertness and tension as they seek reassurance through updates, messages, and online interactions.

Depression and Sadness

Psychological research has identified a clear association between heavy social media use and increased depressive symptoms. One of the primary contributing factors is constant self-comparison with the seemingly perfect lives presented online. Social media platforms often showcase highly curated and idealized moments, as users tend to share only their happiest experiences and greatest achievements. Repeated exposure to these images can lead individuals to feel inadequate or dissatisfied with their own lives. Over time, such comparisons may generate frustration, jealousy, and bitterness, particularly when users measure their personal success against unrealistic online portrayals. When individuals fail to recognize that these images do not reflect full reality, prolonged exposure can gradually shift their emotional state toward sadness and depression.

Lower Self-Esteem and Body Image Issues

Social media platforms are saturated with edited, filtered, and polished images, creating fertile ground for constant self-comparison. Individuals with perfectionist tendencies or high self-expectations may be especially vulnerable to declines in self-esteem when comparing their appearance, lifestyle, or achievements to those of others online. Young people, including both teenage girls and boys, are particularly at risk, as they may feel less attractive or less successful when comparing their ordinary daily lives to influencers or celebrities. Mental health specialists have noted that repeated exposure to idealized images can generate feelings of envy, pain, and self-contempt, reinforcing negative thought patterns and undermining self-confidence. In some cases, this decline in self-esteem is linked to body image disturbances, especially among adolescent girls, due to prolonged exposure to unrealistic beauty standards.

Sleep Disorders and Insomnia

Sleep is often one of the first areas affected by excessive technology use. Prolonged screen exposure before bedtime can disrupt natural sleep patterns and contribute to insomnia. While blue light emitted by screens is often cited as a cause—due to its potential to suppress melatonin production—some experts suggest that its effect may be less significant than commonly believed. Instead, the primary issue often lies in mental stimulation late at night. Engaging with notifications, social media, or video games keeps the brain in an active state, making relaxation difficult. Repeatedly waking up to check the phone further reduces sleep quality. Research indicates that sleep deprivation caused by nighttime screen use leads to daytime fatigue, increased stress, and low mood, particularly among adolescents. This can create a vicious cycle: poor sleep reduces energy for offline activities, which in turn encourages even more screen use.

Digital Addiction and Weakened Self-Control

Repeated and intense engagement with digital devices can develop into psychological dependence and behavioral addiction in some individuals. Specialists define digital or internet addiction as compulsive online use accompanied by extreme difficulty stopping, despite clear negative consequences. This addiction becomes evident when essential life activities—such as family interaction, academic responsibilities, or professional duties—are neglected in favor of constant connectivity. Digitally addicted individuals may gradually withdraw from real-life social engagement, allowing their online world to replace meaningful offline experiences. The issue is intensified by platform designs that rely on endless scrolling, instant rewards, and continuous notifications, all of which make self-control increasingly difficult.

This addictive pattern is often accompanied by emotional distress when attempting to disconnect. Some individuals experience agitation, anxiety, or depressive feelings when separated from their devices, followed by immediate psychological relief upon reconnecting—similar to a temporary emotional "dose." Over time, this cycle reinforces dependence. In the long term, digital addiction may lead to strained family relationships, social withdrawal, and physical complaints such as fatigue, insomnia, and musculoskeletal pain resulting from prolonged device use.

Social Isolation and Feelings of Loneliness

Although technology was designed to enhance communication and connection, its unhealthy use can paradoxically result in social isolation and loneliness. Introverted or socially anxious individuals may find online communication more comfortable than face-to-face interaction. However, relying primarily on virtual relationships can limit opportunities to develop real-life social skills and self-confidence. Over time, this reliance may weaken an individual's ability to engage meaningfully with others offline.

In more severe cases, the virtual world begins to take precedence over real life. Individuals may become less attentive to the people around them and more absorbed in digital interactions, replacing genuine emotional experiences with brief dopamine-driven online rewards. Despite constant online presence, they may experience deep emotional loneliness, as virtual interactions cannot fully replace the emotional fulfillment derived from meaningful real-life relationships and activities. This sense of isolation, despite continuous connectivity, is strongly associated with increased risks of anxiety and depression.

In summary, excessive technology use without clear boundaries can significantly affect mood and psychological wellness. Numerous studies have demonstrated a strong association between heavy screen use and poorer mental health outcomes across different age groups. While technology undoubtedly offers enjoyment, convenience, and valuable benefits, awareness of its psychological risks is essential. By understanding these effects, individuals can take informed steps to regulate their technology use and protect themselves from long-term mental and emotional difficulties.

Social Impacts: Family Relationships, Friendships, and Marriage in the Digital Age

Technology does not influence individuals in isolation; it reshapes the very fabric of social relationships. Smartphones, social media platforms, and constant connectivity have introduced new patterns of interaction within families, friendships, and married life. While technology has made communication faster and more accessible, it has also altered how people relate to one another emotionally and socially. Below is an in-depth analysis of how digital technologies influence different forms of human relationships.

Family Relationships and Raising Children

Modern technologies have introduced significant challenges into homes and family life. Instead of family members gathering in the living room to talk, share stories, or watch a program together, it has become increasingly common for each person to retreat to a separate corner, absorbed in a phone or tablet. As a result, direct communication declines, spontaneous conversations disappear, and emotional warmth within the household gradually fades, creating subtle yet growing emotional distance.

Many parents, consciously or unconsciously, hand digital devices to their children as a quick solution to calm them, occupy them, or gain a moment of quiet. While this may seem harmless in the short term, the long-term consequences can be profound. The child becomes absorbed in the screen and disengages from parental interaction. Over time, this pattern can weaken the emotional bond between parents and children. Parents who are constantly distracted by work messages or social media may unintentionally reduce meaningful interaction with their children, while children may begin to view devices as a primary source of comfort, entertainment, and companionship. This dynamic can foster isolation and emotional detachment from a very early age.

Today, many children turn to the internet to ask their questions instead of engaging in conversation with their parents. Search engines and video platforms increasingly replace family dialogue as sources of information. This phenomenon, often described as "knowledge substitution," reduces opportunities for discussion, shared learning, and values transmission between generations. The loss of dialogue can widen generational gaps and increase misunderstanding and conflict. Parenting experts warn that when this pattern becomes the norm, it may erode traditional family communication values, turning family members into isolated "islands" living together under one roof but emotionally disconnected.

It is also important to note that parental distraction caused by excessive phone use can trigger attention-seeking behavior in children. Some children become restless, irritable, or overly demanding simply because they are seeking their parents' focus. In this sense, technology acts as a double-edged sword within families. On one hand, it enables video calls that bring distant relatives closer; on the other hand, when misused, it pushes those who are physically closest further apart, threatening the warmth, intimacy, and stability of family life.

The Impact of Technology on Friendships and Social Interaction

Social media platforms and messaging applications have made it possible to stay connected with dozens, or even hundreds, of friends and acquaintances at the same time. At first glance, it appears that friendships have expanded across borders and cultures. However, this expansion raises an important question: has the depth and quality of friendship grown as well?

On the positive side, technology offers opportunities for socially shy or introverted individuals to find online communities that share their interests, providing a sense of belonging and emotional comfort. These connections can reduce feelings of loneliness and help individuals feel understood. However, online friendships often lack the emotional depth and intimacy that develop through real-world interaction. Many young people are attracted to the ease and excitement of virtual friendships, only to realize how fragile these connections can be when faced with disagreement, conflict, or life challenges. Communication through screens can reduce empathy, as the absence of facial expressions, tone of voice, and physical presence makes it easier to ignore emotions or abruptly end relationships.

Cyberbullying has also emerged as a serious social issue among peers. Some young people experience ridicule, harassment, or exclusion through social platforms, often from classmates or acquaintances. This digital aggression can cause deep emotional pain, damage trust, and lead to social withdrawal. In addition, misunderstandings spread easily online, as written messages lack nonverbal cues and can be interpreted in unintended ways. Privacy, too, has taken on a new meaning. Personal moments are now publicly displayed, which can create tension among friends—for example, when someone feels hurt after seeing photos of a gathering they were not invited to.

In short, while social media has made it easier to stay in touch despite physical distance, it has not guaranteed strong or meaningful friendships in real life. Research suggests that an increase in digital communication does not necessarily translate into closer relationships offline. Face-to-face meetings, shared experiences, empathy, and active listening remain the true foundations of trust and lasting friendship. Therefore, young people are encouraged to view social platforms as tools that support communication, not as replacements for real human connection.

Married Life in the Age of Social Media

Marriage is among the relationships most affected by the deep integration of technology into daily life. While some couples use digital tools positively—to coordinate responsibilities, share affectionate messages, or stay connected during physical separation—the negative impact is increasingly visible.

One of the most common challenges is digital addiction affecting one or both partners. A husband or wife may spend long hours on a phone, scrolling or interacting online, at the expense of meaningful conversation and emotional presence. This constant distraction reduces attentiveness to a partner's feelings and needs, creating emotional neglect and paving the way for distance and dissatisfaction.

Recent studies indicate a noticeable rise in marital conflict and divorce counseling related to social media use. Some research reports a dramatic increase in relationship disputes linked directly to online behavior. Jealousy and suspicion are among the most common triggers. One partner may monitor the other's activity, misinterpret harmless interactions, or feel threatened by online communication with others. Even small actions, such as liking a photo or replying to a message, can escalate into conflict when trust is fragile.

Comparison is another silent threat to marital stability. Exposure to idealized portrayals of other couples' lives can lead one spouse to feel dissatisfied with their own relationship. Comparing real life, with its responsibilities and challenges, to carefully curated online images can generate frustration, resentment, and unrealistic expectations. Over time, this dissatisfaction may turn into chronic tension and emotional withdrawal.

In some cases, social platforms open the door to inappropriate relationships when boundaries are not respected, undermining trust. In others, both partners retreat into the digital world as a form of escape, avoiding unresolved issues through online distraction rather than communication. Even older couples are not immune. Those unfamiliar with digital environments may fall victim to deceptive online interactions that offer false emotional validation, threatening family stability.

Overall, digital technology has become a new source of pressure on the institution of marriage. This reality highlights the need for shared awareness, mutual respect, and clear boundaries. When used wisely, technology can strengthen connection; when misused, it can quietly erode intimacy and trust.

Adolescents in the Crosshairs of Technology: The Most Psychologically and Socially Affected Group

Although technology affects all age groups, adolescents are widely recognized as the most vulnerable and deeply influenced demographic, both psychologically and socially.

Adolescence is a period marked by identity formation, heightened sensitivity to self-evaluation, a strong desire for peer approval, and susceptibility to trends and social influence. In this context, smartphones and social media become powerful forces shaping emotional wellness. Below are some of the most pressing challenges adolescents face in the digital age.

A Spiral of Self-Comparison and Declining Self-Confidence

Many adolescents spend hours following celebrities, influencers, and peers who present their lives in a constant state of perfection. This continuous exposure to polished images and success stories pushes teenagers to compare their ordinary daily experiences with others' highlight moments. Over time, such comparisons can create the belief that their own lives are less exciting, less successful, or less valuable. Gradually, repeated self-comparison undermines self-esteem.

Psychologists emphasize that many teenagers' emotional states have become closely tied to online feedback, such as likes, comments, and shares. If a post receives limited engagement, a teenager may interpret it as personal rejection or social failure. Seeing peers receive high levels of approval can trigger anxious questions like, "What is wrong with me?" or "Why am I not enough?" This constant competition for digital validation places immense psychological pressure on adolescents.

A recent study in the United Kingdom found that adolescents with mental health difficulties spend approximately 50 additional minutes per day on social media compared with their healthier peers. This finding suggests that some teenagers turn to digital platforms as an escape from emotional distress, even though this escape may intensify their struggles. When self-worth becomes dependent on online validation, it remains fragile and unstable. Confidence built on likes and followers can collapse quickly, unlike confidence built through real-world achievements, skills, relationships, and personal growth.

Cyberbullying and its Psychological Consequences

Bullying has long been a serious challenge within school environments. However, with the rapid expansion of the internet and social media, this problem has taken on a new and more dangerous form: cyberbullying. Today, bullying no longer occurs only in classrooms or playgrounds; it now unfolds through messages, comments, images, and posts shared online. Unfortunately, many adolescents are exposed to public insults, harassment, and humiliation on social platforms by their peers. Hurtful remarks, the spread of rumors, or the circulation of embarrassing images can travel quickly across digital spaces, leaving a teenager feeling trapped, exposed, and publicly shamed.

In such situations, the psychological impact can be severe. Victims of cyberbullying often experience intense distress, embarrassment, sadness, fear, and anger. Over time, they may begin to withdraw socially, avoid school or online spaces, and lose interest in activities they once enjoyed. Mental health specialists agree that cyberbullying significantly increases feelings of loneliness and isolation while weakening adolescents' sense of emotional and psychological safety. In some tragic cases reported around the world, persistent online bullying has pushed certain adolescents toward self-harm or suicidal thoughts, as they struggle to find a way to escape the overwhelming pressure.

One of the most dangerous aspects of cyberbullying is its constant presence. In the past, when the school day ended, a bullied student could return home and find a sense of safety and relief. Today, however, bullying can follow the victim into their bedroom through a mobile phone, continuing day and night without pause. This reality makes parental awareness and involvement absolutely essential. Parents need to remain alert to sudden changes in a child's mood, behavior, or sleep patterns, which may indicate online harassment. Encouraging open dialogue, offering emotional support, and seeking professional or school-based help can play a crucial role in protecting adolescents from further harm.

At the same time, schools and local communities share responsibility in addressing this issue. Educational institutions must establish clear digital conduct policies, accessible reporting mechanisms, and firm deterrent measures to reduce cyberbullying in online environments. Without collective action, this hidden form of abuse can continue to grow unchecked.

Loss of Connection to Reality and Disengagement from the Real World

By nature, adolescents are drawn to experiences that offer immediate pleasure, stimulation, and excitement. Modern electronic games, social media platforms, and digital entertainment have successfully provided an endless stream of interaction and instant gratification. However, when used excessively, these technologies can pull teenagers away from real life to a degree that weakens their connection to reality itself.

Today, it is increasingly common to see young people spending long hours alone in their rooms, fully absorbed in screens, while neglecting school responsibilities, physical activity, and hobbies that support healthy development. Psychology specialists warn that many adolescents have begun to prefer phone screens over real-world experiences and the gradual development of genuine talents. For example, instead of learning to play a musical instrument and experiencing the deep satisfaction of progress and achievement, a teenager may simply watch videos of others performing. This provides a quick dopamine boost and short-term pleasure, but it cannot replace the long-lasting sense of fulfillment that comes from real effort and accomplishment.

In this way, technology can create a dangerous cycle: an addiction to fast pleasure without genuine satisfaction. Over time, this pattern may lead to boredom, lethargy, reduced motivation, and even depression, as the teenager is not building real skills or achieving meaningful goals in their life.

Closely related to this issue is the concept of an "alternative reality." Some adolescents retreat into digital environments such as online games, virtual worlds, or internet communities to escape real-life struggles, whether academic pressure, social difficulties, or family conflicts. Within these virtual spaces, they may experience a sense of control, comfort, and achievement that feels absent from their daily lives. However, this prolonged digital immersion makes returning to real-world responsibilities increasingly difficult. The deeper the involvement in virtual life, the weaker the interest in what is actually happening around them.

Evidence of this trend appears in internet addiction treatment camps in South Korea, which will be discussed later. Some adolescent participants have admitted statements such as, "I find staying on my mobile phone more enjoyable than being with my family or friends." This powerful admission highlights how dangerous disconnection from reality can become when a young person prefers the virtual world over real human relationships. For this reason, guiding adolescents toward balance is essential. Technology can certainly be enjoyed, but it must be complemented by real-life engagement through sports, creative activities, face-to-face social interaction, and experiences that provide tangible joy and a genuine sense of accomplishment.

A Generation Under Pressure: Adolescents in the Digital Era

In summary, today's adolescents are living through a historically unprecedented stage of human development. They are growing up surrounded by technology in nearly every aspect of life, from education and entertainment to social interaction and self-expression. This reality places a major responsibility on families, schools, and society as a whole. There is an urgent need to raise awareness about the risks of excessive technology use and to equip teenagers with the psychological tools required to cope with the pressures of the digital age.

These tools include building self-confidence without relying on online validation, encouraging honest communication when problems arise online, and nurturing real-world interests that protect adolescents' mental balance and overall wellness. Without such guidance, young people may struggle to navigate a digital world that offers constant stimulation but limited emotional support.

Examples from the Gulf and the World: Similarities and Differences in Impact

The challenges discussed above are not limited to a single society or culture. They are global phenomena affecting countries with different traditions, values, and levels of technological advancement. Examining examples from the Gulf states and countries such as

Japan, Sweden, and South Korea helps highlight both shared patterns and important cultural differences in how technology shapes mental and social wellness.

The Gulf States

Gulf countries are characterized by exceptionally high rates of smartphone ownership and social media use, particularly among young people. For instance, recent statistics indicate that approximately 94 percent of Saudis use social media, with the average individual spending around three hours per day browsing these platforms. This figure exceeds the global average of roughly two and a half hours. Similarly, the United Arab Emirates consistently ranks among the top countries worldwide in terms of smartphone penetration and daily time spent on social networks.

This intense digital engagement has made the negative effects of excessive technology use increasingly visible in Gulf societies. Families face growing challenges related to children and adolescents becoming deeply attached to electronic games and online content. In response, many schools and educational authorities have launched awareness campaigns focused on digital wellness. In the UAE, for example, public awareness of technology's impact on mental health has risen, with local media frequently sharing advice on achieving digital balance and avoiding harmful overuse.

Mental health professionals in the Gulf have also reported a rise in cases related to digital addiction among young people, noting its impact on attention span, academic performance, and daily functioning. At the same time, concerns about cyberbullying through messaging apps and social media platforms have become more prominent.

Importantly, Gulf governments are also investing in the positive use of technology. Saudi Arabia, for example, has introduced government-supported digital applications designed to strengthen mental health services, including platforms that offer counseling and psychological support, making help more accessible to those in need.

The main similarity between the Gulf and other parts of the world lies in the emergence of technology-related mental health challenges such as anxiety, depression, and social isolation. The key difference, however, is cultural context. Gulf societies place strong emphasis on family bonds and social cohesion, which leads parents to view excessive screen use with particular concern due to its potential impact on family relationships and traditional values. At the same time, this strong family structure can also serve as a protective factor when parents actively guide and regulate children's technology use through shared rules and open communication.

Japan and the Hikikomori Phenomenon: Chronic Social Withdrawal

Japan offers a striking example of how technology intersects with social and psychological challenges. While the country is globally recognized for its technological innovation, it has also faced one of the most extreme social phenomena of the modern era: hikikomori. This term refers to prolonged and severe social withdrawal and is used to describe more than one million young people in Japan who have chosen near-total isolation from society.

Hikikomori individuals may remain confined to their rooms for months or even years, avoiding school, work, and social contact altogether. This phenomenon has complex causes, including academic pressure, economic stress, and cultural expectations, and it cannot be attributed to technology alone. However, the availability of the internet, online gaming, and digital services has enabled some socially withdrawn individuals to survive without leaving their rooms. Through apps, they can order food, communicate minimally via text, and fill their time with gaming or video streaming, all while avoiding direct human interaction.

This example demonstrates how technology, while not the sole cause of social withdrawal, can unintentionally support and prolong isolation when deeper psychological and social issues remain unresolved. A striking paradox emerges at this point. Contrary to what one might intuitively expect, Japanese government studies have revealed that many individuals classified as hikikomori do not necessarily spend more time online than their peers. In other words, some experience profound emptiness, loneliness, and social isolation without heavy engagement in digital activities. Nevertheless, the availability of modern technology undeniably makes it easier for individuals to remain isolated for extended periods, far longer than would have been possible in earlier times when basic daily needs could not be met from within the home. Technology, while not always the root cause, can therefore act as a silent enabler of prolonged withdrawal.

In response to this growing crisis, Japanese society has made sustained efforts to address the problem through specialized support centers and gradual reintegration programs. These initiatives aim to help hikikomori individuals slowly reconnect with education, work, and everyday social life at a pace that respects their psychological condition. The key lesson emerging from Japan's experience is that excessive technology use is not always the primary cause of isolation; in many cases, it is a symptom of deeper psychological, emotional, or social struggles. At the same time, Japan clearly recognizes the importance of balance. Schools have introduced programs that teach students healthier technology habits, and many families impose strict limits on gaming time and screen use in an effort to reduce the risk factors that may contribute to withdrawal and social disengagement.

In terms of similarities with other countries, Japan shares the global trend of increasing youth isolation and reduced face-to-face interaction, trends that are partly supported by digital alternatives. The difference, however, may lie in cultural context. Japanese culture traditionally places high value on privacy, restraint, and emotional self-control, factors that may have created conditions in which an extreme form of withdrawal could develop on a scale not commonly seen in many other societies.

Sweden and the Nordic Countries

Sweden provides another instructive case. It is among the countries with the highest rates of internet and smartphone adoption in Europe. At the same time, Swedish society is often recognized for its strong public health awareness and sense of collective responsibility. Swedish adolescents, like their peers elsewhere, spend many hours each day on screens. A large-scale study involving thousands of teenagers found that the majority of Swedish youth use social media for more than three hours per day. Alongside this trend, similar mental health concerns have been observed. Swedish research has linked prolonged social media use to poorer sleep quality, increased depressive symptoms, and heightened emotional distress among adolescents. Other studies have also documented negative effects on body image resulting from repeated exposure to idealized and unrealistic online content.

What distinguishes the Swedish experience, however, is the way the country responds to these challenges. Rather than framing technology as an enemy, Sweden has focused on prevention, balance, and digital health innovation. Online mental health counseling services are widely available and socially accepted, offering young people a familiar, private, and accessible channel for seeking help when they feel overwhelmed or distressed. In addition, the Swedish education system integrates digital literacy into school curricula, teaching children and adolescents how to manage screen time, protect their privacy, and interact respectfully in online environments.

More broadly, Swedish society adopts a balanced and pragmatic approach. The country is a leader in using technology to improve the quality of life through telehealth services, educational platforms, and digital public services, while openly acknowledging the risks of overuse. The government actively funds research to better understand the relationship between screen time and mental health, and it supports awareness campaigns that promote physical activity, outdoor engagement, and reduced sedentary behavior.

In short, Sweden faces the same global challenges, including anxiety, depression, sleep disruption, and body image pressure among young people. The difference lies in its advanced, community-based strategy for responding constructively. Rather than relying solely on restrictions or bans, Sweden emphasizes guidance, education, and support, encouraging young people to use technology as a helpful tool for learning, mindfulness, and health management without allowing it to dominate their lives. This approach offers a valuable model for other societies seeking to strengthen digital wellness.

South Korea and the Challenge of Internet Addiction

South Korea represents one of the most technologically connected societies in the world. Its advanced digital infrastructure enables high-speed internet access and seamless video streaming even in public spaces such as metro stations. Alongside this deep integration of technology into daily life, however, the country has faced notably high rates of internet and online gaming addiction among young people. Surveys estimate that approximately 10 percent of Korean adolescents experience problematic internet use to varying degrees, a figure that has raised serious concern among health professionals and policymakers.

What makes South Korea's case particularly significant is the early recognition of the problem and the strong national response. For several years, the government enforced what became known as the "Cinderella Law," which restricted individuals under the age of 16 from accessing online games after midnight, despite the fact that some teenagers found ways to bypass the restriction. Beyond regulation, South Korea invested heavily in treatment and rehabilitation. Specialized counseling centers and internet addiction camps were established for adolescents assessed as experiencing digital addiction.

These rehabilitation programs, typically lasting three to four weeks, involve complete digital separation. Participants surrender their devices upon arrival and follow structured daily routines focused on physical exercise, face-to-face social interaction, psychological counseling, and the rebuilding of healthy habits. During the program, adolescents learn stress management techniques and participate in activities such as hiking, rock climbing, music lessons, and team-based challenges. The central aim is to reconnect young people with real-life pleasure, achievement, and self-discipline away from screens.

The voices of the participants themselves are particularly revealing. In post-camp evaluation surveys, many adolescents admitted that they had previously lied about the number of hours they spent online and that they found digital devices more enjoyable than spending time with family or friends. These honest confessions illustrate how deeply digital addiction can widen the gap between adolescents and their real social environments.

South Korea's experience reflects a global similarity in the spread of digital addiction, but it also stands out for the seriousness and scale of its response. Few countries have taken such decisive steps by establishing specialized treatment facilities and framing digital addiction as a public health issue. As a result, South Korea is often seen as a pioneer in group-based rehabilitation strategies for adolescents, driven by a broader awareness that the mental wellness of future generations is at risk if left unaddressed.

Toward a Digital Balance that Supports Psychological and Social Wellness

After this extensive review of the psychological and social effects of modern technology, one conclusion becomes unmistakably clear: technology is a double-edged sword. On the one hand, its benefits to education, work, and communication are undeniable. Technology has connected distant people, democratized access to knowledge, and enabled communication across borders and cultures. On the other hand, this discussion has shown how unbalanced and excessive use can harm mental health and weaken social bonds. Conditions such as anxiety and depression may intensify, and human relationships can deteriorate despite the abundance of communication tools. Social media, in particular, has created a parallel virtual world that offers real advantages alongside unprecedented challenges.

As a result, digital wellness has emerged as a central concept of our time. Digital wellness refers to achieving a healthy and intentional balance between online engagement and offline life. Once we understand the psychological and social consequences of excessive use, including sleep disruption, cyberbullying, emotional isolation, and family fragmentation, the need for public awareness becomes clear. This responsibility is shared. Families play a crucial role in guiding children and modeling mindful technology use. Schools and educational institutions must integrate digital media literacy, online safety, and responsible behavior into their curricula. Governments also have an important role to play through national awareness campaigns, mental health services, and support programs for individuals affected by technology-related challenges.

At the heart of this discussion lies the principle of moderation. Technology is not an enemy to be eliminated, but a tool that should serve human needs rather than dominate human lives. Just as some individuals escape reality through screens, others use technology to enhance their lives by learning new skills, maintaining family connections through video calls, or benefiting from mindfulness and mental health applications. Awareness and balance are therefore essential. Learning when to engage with technology and when to step away allows individuals to be fully present with those around them. Experts consistently recommend setting device-free times, particularly during family interactions and before sleep, as these simple habits can significantly improve emotional comfort and social connection.

Finally, building a social culture that speaks openly about issues such as digital depression and digital addiction can reduce stigma and encourage those affected to seek help without shame. Digital mental health must be recognized as an integral part of public health. Encouragingly, signs of this awareness are already emerging. In the Gulf, institutions have begun hosting seminars on protecting mental health in the age of social media. Globally, governments are introducing legislation to protect children online, and technology companies are adding features that allow users to monitor and limit screen time.

In conclusion, digital wellness is a shared responsibility. Achieving it leads to healthier individuals, stronger family relationships, and a more conscious and productive society. Technology was created to improve human life, not to become a source of stress and disconnection. Let balance guide our approach. By combining awareness, self-discipline, and supportive policies, societies can benefit from the digital age while preserving mental health, human connection, and a sense of shared humanity for future generations.

Chapter Five:

"Digital Privacy and the Protection of Personal Data"

The Concept of Digital Privacy and Its Importance in Everyday Life

Digital privacy refers to an individual's ability to protect personal information and to control who can access it within the online environment. It encompasses everything related to our data, identity, and personal activities when we use digital technologies and the internet.

In modern daily life, people rely heavily on smart devices and digital services for communication, work, education, shopping, and entertainment. As a result, an enormous volume of personal data flows continuously through online systems. For this reason, digital privacy has become increasingly important, as it aims to protect this information from unauthorized access, misuse, or exploitation. Digital privacy can be understood as an extension of traditional privacy into the digital age, where every individual holds a fundamental right to keep their personal information private, secure, and under their control.

For example, when we share family photographs, communicate sensitive health information through messaging applications, or store personal documents on cloud-based services, we naturally expect these materials to remain private and protected. Safeguarding digital privacy provides a sense of security and confidence while navigating the online world, and it shields individuals from serious risks such as identity theft, blackmail, data breaches, and violations of personal dignity.

The importance of digital privacy lies in its role in protecting the most sensitive aspects of our lives within a fast-moving and interconnected world. When individuals secure their digital privacy, they are not only protecting abstract data, but also safeguarding their family life, financial stability, personal reputation, and health information. Respect for digital privacy also builds trust in online environments. Whether conducting financial transactions, shopping online, or communicating with others, users feel more confident when they know their information will not be misused or exploited without their consent.

Moreover, digital privacy supports personal freedom. When individuals feel that they are not constantly monitored or surveilled, they are more likely to express themselves openly and interact online without fear. From this perspective, digital privacy is a foundational condition for digital wellness in everyday life and a core pillar for protecting digital rights in an era where technology is deeply embedded in even the smallest details of human activity.

Types of Personal Data Collected Digitally About Users

To understand digital privacy risks, it is essential to recognize the wide range of personal data that can be collected when individuals use digital services. Every time a person connects to the internet or interacts with applications, multiple types of data may be gathered, stored, and analyzed. The most common categories include the following:

- **Basic identifying information:** This includes data such as name, home address, email address, phone number, and official identification numbers (for example, national ID or passport numbers). Users usually provide this information voluntarily when creating accounts, registering for services, or completing online forms.

- **Financial and transaction data:** This category includes credit card details, bank account information, and records of online purchases. Such data is extremely sensitive, as unauthorized access can lead directly to financial fraud, theft, or serious economic harm.

- **Geolocation data:** Many applications and devices collect location information through GPS, Wi-Fi networks, or IP addresses. Smartphones, for instance, can track a user's location in real time, and apps may use this data to offer services such as navigation, traffic updates, or nearby recommendations. However, location data can also reveal detailed patterns about daily routines, habits, and personal movements.

- **Browsing and search history:** Every website visited and every search query entered may be recorded by browsers, search engines, or third-party services. Over time, this data creates a detailed digital record of interests, habits, opinions, and intellectual tendencies, forming a personal profile that reflects an individual's online behavior.

- **Health and biometric data:** With the widespread use of wearable devices and health-related applications, data may be collected about physical activity, heart rate, sleep patterns, and overall health status. Biometric data also includes fingerprints, facial recognition data, and iris scans used for authentication. This information is among the most sensitive, as it relates directly to both physical identity and medical conditions.

- **Images, videos, and audio recordings:** Personal photographs, family videos, voice messages, and audio recordings captured by digital assistants all constitute personal data. Although these may appear harmless or routine, they can reveal identities, relationships, locations, and daily activities.

- **Behavior and interaction patterns:** Many digital services collect data about how users interact with platforms, including time spent on apps, types of content viewed, posts liked or shared, and times of peak activity. In some cases, even typing speed or touch patterns are analyzed. This behavioral data helps companies build detailed user profiles.

- **Device and connection identifiers:** Technical data such as IP addresses, device identification numbers, operating system details, and browser information is often collected automatically. These identifiers help distinguish one user or device from another and are frequently used for tracking and analytics purposes.

In general, any information that relates to an identifiable individual is considered personal data. Some forms of data are obvious and visible, while others operate silently in the background. Understanding these categories allows individuals to better recognize how privacy risks arise and why proactive protection is necessary.

How Do Apps and Websites Collect This Data, and Why?

When users interact with apps and websites, various technologies are employed to collect personal data, often without clear awareness on the part of the user. One of the most common tools used for this purpose is cookies.

Cookies are small text files stored on a user's device when visiting a website. They allow the website to recognize the browser, remember login details, and store user preferences for future visits. Another common tool is tracking pixels, which are tiny invisible images embedded in webpages or emails. When these pixels load, they notify the sender that a page has been viewed or an email has been opened.

Some platforms also rely on device fingerprinting, a technique that collects a combination of device and browser characteristics, such as screen size, system settings, and installed fonts or extensions. This information is used to create a unique "fingerprint" that allows tracking even when cookies are disabled.

Mobile applications often request permissions to access features such as location services, contacts, microphones, or photo galleries. When users grant these permissions, the app can collect and transmit that data to its servers. In addition, nearly every interaction, clicks, scrolling behavior, navigation paths, and time spent on screens, can be logged and analyzed.

This raises an important question: why do companies collect such extensive data? The reasons vary, and they range from legitimate purposes to purely commercial motives.

- **Personalization and service improvement:** Websites may store user preferences, such as language or display settings, to improve convenience. Companies also analyze user behavior to refine design, functionality, and overall user experience.

- **Delivering relevant content and features:** Location data enables services like local news, weather updates, or nearby recommendations. Browsing history helps algorithms suggest content aligned with user interests, such as educational videos or articles.

- **Analytics and market research:** Aggregated data allows companies to identify trends and patterns across large user populations, helping them decide which features to develop or discontinue.

- **Targeted advertising and profit generation:** Many platforms collect data to create detailed advertising profiles. Personalized ads are more effective, increasing revenue for companies. This commercial motive is one of the strongest drivers of data collection today.

- **Security and fraud prevention:** Certain data is collected to protect users, such as monitoring unusual login activity or detecting suspicious behavior that may indicate hacking or fraud.

- **Sharing data with third parties:** In some cases, companies share or sell user data to advertisers, analytics firms, or research organizations. Third-party tracking tools embedded in websites can monitor user behavior across multiple platforms, creating extensive cross-site profiles.

In simple terms, companies collect data to understand users better. Sometimes this understanding is used to enhance services, and sometimes it is used primarily to monetize user behavior. For this reason, individuals must be aware of how their data is collected and used. Awareness empowers users to enjoy digital services while maintaining reasonable control over their privacy and personal information.

How Users Can Protect Their Data Online

Despite the many privacy challenges that exist in today's digital environment, there are several practical and effective steps that users can take to protect their personal data and reduce potential risks. By adopting simple habits and making use of available settings, individuals can significantly strengthen their digital privacy. The following best practices are widely recommended for everyday online use and can be applied by users of all ages and technical backgrounds.

1.   Review and adjust privacy settings regularly:

Users should develop the habit of regularly reviewing the privacy settings of their online accounts and mobile applications. Most digital platforms, including social media networks and mobile apps, provide privacy controls that allow users to decide what information is visible to others and what remains private. For instance, personal details such as email addresses and phone numbers should not be publicly displayed on profiles. In addition, users should review app permissions on their devices and question whether each permission is truly necessary. Why would a simple game need access to photos, contacts, or location data? Unnecessary permissions should be disabled immediately. Browsers also include privacy settings that allow users to block location tracking and limit unauthorized data access. These tools represent the first line of defense, and configuring them according to one's desired privacy level is essential.

2.   Share less and think carefully before posting:

Any personal information shared online should be treated as sensitive by default. Before publishing a post, users should ask themselves whether they would feel comfortable if strangers accessed this content. It is advisable to avoid sharing highly sensitive information, such as home addresses, national identification numbers, private family matters, or detailed travel plans. Online content can spread far beyond its intended audience and often remains accessible for long periods of time. Similarly, when registering for digital services, users should avoid providing optional information unless it is absolutely necessary. The less data that is shared, the lower the risk of misuse or leakage. Since the internet rarely forgets, managing one's digital footprint intentionally is a key aspect of privacy protection.

3. Use strong and unique passwords for every account:

Passwords act as the primary gateway to online accounts, making their strength critically important. Users should create long passwords of at least twelve characters that combine letters, numbers, and symbols. Easily guessed information, such as names or dates of birth, should be avoided. Most importantly, the same password should never be reused across multiple platforms. If one account is compromised, reused passwords can expose many others. Because remembering multiple complex passwords can be challenging, using a password manager is strongly recommended. Password managers can generate strong passwords and store them securely in encrypted form, protected by a single master password. Tools such as Bitwarden and KeePass reduce the mental burden on users while greatly enhancing overall security.

4. Enable two-factor authentication:

Adding an additional layer of verification during login has become essential in the modern digital landscape. Even if a password is exposed during a data breach, two-factor authentication can prevent unauthorized access by requiring a second form of verification. Users should enable this feature wherever it is available. Typically, this involves entering a temporary code sent to a phone, email, or generated by an authenticator application such as Google Authenticator. This extra step significantly strengthens account security and helps protect personal data even if login credentials are compromised.

5.   Use encryption to protect sensitive information:

Encryption is one of the strongest tools available for safeguarding privacy. It converts data into an unreadable format that can only be accessed by someone with the correct decryption key. Users should prioritize services that integrate encryption by design. For example, messaging applications that offer end-to-end encryption ensure that even the service provider cannot read message content. Signal is a well-known example. For email communication, encrypted email services such as Proton Mail offer enhanced privacy compared to traditional providers. Users can also encrypt sensitive files before uploading them to cloud storage or sharing them online. In addition, enabling full device encryption on smartphones and computers ensures that locally stored data remains protected if the device is lost or stolen.

6.   Be cautious with suspicious links and messages and remain alert to phishing attempts:

A large number of security breaches begin with deceptive messages delivered through email, text messages, or social media platforms.

Phishing involves tricking users into revealing sensitive information by pretending to be a trusted organization. Any message that requests personal details or directs users to log in through a link should be treated with caution. Users should avoid clicking on unverified links and always check website addresses carefully before entering login credentials. For example, banks and official institutions rarely request sensitive information through email. Security software and browser extensions that scan links and attachments can provide additional protection. When in doubt, users should ignore the message or contact the organization directly through official channels.

7. Use security software and keep all systems updated:

Protecting personal data also requires securing the devices used to access it. Installing a trusted and regularly updated antivirus program on computers and mobile devices helps detect malicious software such as spyware and keyloggers. Operating systems and applications should always be updated to their latest versions, as updates frequently include security patches that fix known vulnerabilities. Ignoring updates can leave devices exposed to attackers who exploit outdated software. Additionally, users should avoid installing applications from unofficial sources, as these may contain hidden malware. In short, strong device security is closely linked to strong digital privacy.

8. Control your digital footprint and review privacy policies:

Every online service has a privacy policy explaining what data is collected and how it is used. Although these documents can be lengthy, users should at least review the key points, such as whether data is shared with third parties and how long it is stored. Several tools and websites now provide simplified summaries of privacy policies to make them easier to understand. Users also have legal rights in many countries to request access to their data or ask for its deletion. Many platforms allow users to download their stored data or permanently delete accounts. Exercising these rights helps users understand what information companies hold about them and restores a sense of control over personal data.

Together, these practices form a protective shield for digital privacy. Users do not need to implement all measures at once. Starting with strong passwords and two-factor authentication, then gradually improving other settings and habits, can make the process manageable. With awareness and consistent effort, individuals can enjoy the benefits of the digital world while keeping their personal data as secure as possible.

The Difference Between Digital Privacy and Cybersecurity

Digital privacy and cybersecurity are often used interchangeably, but they are not identical concepts. While they are closely connected, each has a distinct focus and scope. Understanding the difference between them helps clarify how personal data is protected in the digital world.

Digital privacy:

Digital privacy concerns the protection of an individual's personal information and the right to control how that information is collected, used, and shared. It emphasizes consent, transparency, and confidentiality, ensuring that personal data is not accessed or exploited without permission. Privacy focuses on the ethical and individual dimension of data use. For example, digital privacy ensures that personal photos, messages, location data, and health records remain within boundaries defined by the user.

The field of digital privacy is guided by principles and policies that regulate data collection and processing, including the right to access, correct, or delete personal information. In essence, digital privacy seeks to protect individuals from unwanted intrusion and misuse of their personal data. It is primarily concerned with private, identifiable information rather than public or institutional data.

Cybersecurity:

Cybersecurity refers to the protection of digital systems, networks, and information from cyberattacks, breaches, and disruptions. Its primary goals are to maintain data confidentiality, preserve data integrity, and ensure system availability. Cybersecurity focuses on preventing unauthorized access and defending against malicious actors who seek to steal, damage, or manipulate information.

This field relies on technical tools and measures such as firewalls, intrusion detection systems, encryption technologies, and secure network architectures. For example, securing a website against hacking attempts or protecting a corporate network from ransomware attacks are core cybersecurity tasks. Unlike privacy, cybersecurity applies to all types of data, whether personal, corporate, or governmental.

Although their definitions differ, digital privacy and cybersecurity are deeply interconnected. Privacy cannot exist without strong security measures to protect data, and security efforts are most meaningful when they ultimately safeguard people's personal information. Many privacy-focused practices, such as limiting data collection or anonymizing information, also contribute to stronger security.

A useful analogy is that cybersecurity represents the locks on the doors that prevent intruders from entering, while privacy represents the curtains that prevent others from seeing inside. Both are necessary for safety and comfort. Similarly, in the digital world, privacy and security must work together to build trust.

In conclusion, digital privacy focuses on individuals' rights to control and protect their personal data, while cybersecurity concentrates on defending systems and information from technical threats. Privacy is shaped by laws, policies, and ethical norms, whereas cybersecurity relies on technical safeguards and defenses. Despite these differences, they remain two sides of the same coin. As digital risks continue to grow, strengthening both privacy and cybersecurity has become essential to creating a safer and more trustworthy digital environment.

Tools and Apps to Strengthen Digital Privacy

Fortunately, ordinary users no longer have to rely only on personal caution to protect their privacy. Today, many tools and apps are designed specifically to strengthen digital privacy and make it easier to manage. Think of these tools as an extra shield you can use while browsing the internet or communicating through different apps. Below are some of the most useful solutions that can help you regain control over your digital data.

Use privacy-focused search engines: Instead of traditional search engines that track almost everything you search for, try alternative search engines that do not store your search history or sell your data for advertising. Well-known options include DuckDuckGo, Start page, and Qwant. These engines generally avoid identifying you during searches, which means your queries remain more private.

For example, DuckDuckGo offers a browser extension and a mobile app that help block trackers from Google and other parties while you browse. It is also important to note that Incognito Mode in common browsers is not a magic privacy solution, as many people assume. It mainly prevents activity from being saved on your local device, but it does not stop websites, internet service providers, or other entities from tracking you online. Privacy-oriented search engines, like DuckDuckGo, on the other hand, help keep the search process itself more confidential. Hence, choosing a privacy-respecting search engine is an easy and effective step to reduce tracking across the web.

Use a secure browser and privacy protection extensions: Your choice of web browser can make a real difference in your privacy. Some browsers, such as Brave or Firefox, when properly configured, offer built-in tools to block ads and tracking cookies.

You can also strengthen your current browser by installing privacy-focused extensions, such as Privacy Badger from the Electronic Frontier Foundation, or unblock Origin, to block ads and trackers. These extensions prevent many external parties from monitoring your movement across websites. It is also recommended to enable the "Do Not Track" setting in your browser. Even though websites are not forced to honor it, it still signals your preference not to be tracked.

Finally, do not forget to keep your browser updated so you benefit from the latest security and privacy fixes.

Virtual Private Network services, VPN:

A VPN is a service that creates an encrypted connection between your device and a private server, and then you browse the internet through that server. This hides your real IP address, so your browsing appears to come from the VPN server location rather than your actual location.

The benefit is an extra layer of privacy that can help keep your online activity away from the view of your local internet service provider or anyone trying to snoop on the network. For example, when you use public wireless internet in places like cafes or airports, your connection can be exposed to interception. A VPN encrypts what you send and receive, making it much harder for others to capture or read.

That said, you should choose a trustworthy VPN provider with a strong reputation, because the VPN provider itself could potentially see your connection traffic. Look for services with strict no-logs policies and clear commitments not to share data. Examples often recommended by privacy experts include Mullvad and ProtonVPN. And of course, remember that a VPN is meant to protect privacy, not to bypass laws or engage in illegal activity.

Password Managers:

We have already discussed how important it is to use strong and unique passwords, and this is exactly where tools like password managers become practical and extremely useful. A password manager is an app that stores all your passwords in an encrypted digital vault, so no one can read them except you, using a master password. It also generates highly complex random passwords for every account.

The result is double protection: passwords that are extremely difficult to crack, and no password reuse across different websites and services. Some of the most popular password managers include

1Password, LastPass, and Bitwarden, which is open source and offers a free plan, among many others.

With a password manager, you do not need to remember more than one password: the master password. The app will securely fill your strong passwords for you across websites and apps. Just make sure you choose a trusted manager, create a very strong master password, and enable two-factor authentication on the manager itself if available.

Secure Messaging and Privacy-Friendly Alternatives:

In a world where a few messaging and social apps dominate billions of users, it is worth considering alternatives that respect your privacy more. For example, WhatsApp uses end-to-end encryption for messages, but it still collects a considerable amount of metadata about users, and it is owned by Meta, a company widely known for extensive data collection.

Suggested alternatives include Signal, an open-source messaging app that focuses heavily on privacy, collects minimal personal data, and does not show ads. Another option is Telegram, which offers Secret Chats with strong encryption. However, it is important to note that regular Telegram chats are not end-to-end encrypted by default. Other options include Threema and Wire.

For email, you can use services like Proton Mail or Tutanota, which encrypt your messages and generally require less personal information during sign-up. Switching platforms may take time, and you might need to convince friends or colleagues to use them with you, but when sensitive information is involved, or if you value privacy highly, these alternatives are worth considering.

Additional Tools to Strengthen Privacy

Several other solutions can help keep your identity and online activity more private.

One example is email alias services, which give you disposable or alternate email addresses to use for sign-ups instead of your real email, reducing spam and limiting tracking. Services like Firefox Relay provide this feature easily.

You can also use the Tor Browser if you need a higher level of anonymity. Tor routes your connection through multiple servers around the world and hides your IP address, making tracking far more difficult. Keep in mind that Tor can be slower, and it is typically used for anonymous browsing rather than everyday mainstream use.

There are also privacy-focused operating systems and phones. For example, GrapheneOS on compatible phones aims to reduce tracking components commonly tied to large tech ecosystems. These solutions are more advanced, but knowing they exist can be helpful if you ever need stronger privacy measures.

By using these tools and apps, you increase your ability to control your data and narrow down who can track you or collect information about you. Many of them are free or low-cost, so it is usually more about awareness and habit-building than major spending.

Try one change at a time: use a privacy-focused search engine for a week, or test a secure messaging app with your family, and you will start noticing the difference gradually. Strengthening your digital privacy is easier than ever today, so do not hesitate to equip yourself with these tools.

Privacy Laws and Data Breaches, Local and Global Examples

Many countries and governments around the world have recognized how serious privacy violations have become in the digital age, and how necessary it is to establish laws and regulations to protect users' personal data. At the same time, we have witnessed some of the biggest data scandals and breaches in modern history, which accelerated public awareness and pushed regulation forward more quickly. In this section, we will review examples from the Gulf region, as well as global examples, focusing on both applicable laws and major data breach incidents.

Local context, the Gulf states

Gulf governments have taken major steps to strengthen the protection of individuals' privacy in the digital environment. In the United Arab Emirates, the first federal law dedicated to personal data protection was issued as Federal Decree Law No. 45 of 2021 on the Protection of Personal Data. This law provides a comprehensive framework to ensure the confidentiality of information and safeguard individuals' privacy across the country. It defines the rights of individuals, the data subjects, and the obligations of entities that collect or process data, including the requirement to obtain explicit consent from the data subject before processing personal data, with certain exceptions related to public interest and legal procedures. It also grants individuals the right to request the correction or deletion of their data, and it sets rules for cross-border data transfers. This law, enforced in January 2022, is considered a major milestone as the first comprehensive federal law for digital privacy protection in the UAE. In parallel, the UAE Data Office was established as a regulatory body to oversee the law's implementation and to develop national-level data protection policies.

In September 2021, the Kingdom of Saudi Arabia introduced the Personal Data Protection Law through Royal Decree No. M-19, establishing the first comprehensive legal framework of its kind within the Kingdom. The law underwent a review, resulting in amendments to some of its articles in 2023. Nevertheless, its mandatory enforcement commenced on 14 September 2023. To help companies and institutions comply with the law, Saudi authorities provided a transitional period that lasted for one year following its enforceability, concluding in late 2024. The Saudi Data and AI Authority (SDAIA) is responsible for overseeing implementation and ensuring compliance, having issued regulations that include guidelines for data transfer outside the Kingdom and requirements for appointing data protection officers, among other aspects.

Similar to other contemporary data protection frameworks, the Saudi law emphasizes the necessity of prior consent for lawful processing of personal data, with few exceptions, and mandates entities to safeguard data while adhering to principles like data minimization and secure disposal of data once the collection purpose has been fulfilled. These initiatives align with Vision 2030 and the wider digital transformation agenda, aiming to enhance public confidence in the digital economy. Furthermore, they position Saudi Arabia alongside nations that prioritize user privacy.

It's important to acknowledge that other Gulf nations have taken comparable actions. Bahrain enacted its Personal Data Protection Law in 2018, while Qatar issued a data privacy law in 2016, becoming one of the early Gulf states to pursue this area. These laws are primarily influenced by the European General Data Protection Regulation (GDPR), establishing principles such as transparency, consent, and individuals' rights to access and amend their data, alongside penalties for violations. For instance, Bahrain's law allows for fines of up to one million Bahraini dinars in cases of improper use of personal data. This illustrates the increasing regional recognition of privacy protection as a fundamental component of a secure digital environment that fosters investment and builds public trust.

On a global scale, the European Union stands out as a frontrunner, particularly since the GDPR took effect in May 2018. The GDPR is regarded as a pivotal moment worldwide due to its extensive scope and stringent requirements. It applies to any organization processing the data of EU citizens, irrespective of the organization's location, compelling companies globally to upgrade their privacy protocols. The GDPR grants individuals significant rights regarding their data, including the ability to obtain a copy, correct or delete it, object to marketing-related processing, and transfer data between service providers. Additionally, it requires organizations to adhere to principles such as transparency in outlining data collection and usage, minimizing the data gathered to what is absolutely necessary, and implementing the concept of Privacy by Design in system development. The regulation is also noted for its severe penalty structure, which can reach fines of 4 percent of a company's annual global revenue or 20 million euros, whichever is greater. In practice, substantial fines have been levied under GDPR against major tech companies. For instance, in 2023, Irish authorities imposed a record fine of approximately 1.2 billion euros on Meta, the parent company of Facebook, for transferring European data to servers in the United States without adequate privacy protections. This fine was the largest issued under GDPR to date, highlighting the rigorous enforcement of privacy regulations by European authorities. Such measures have also inspired numerous countries to establish similar laws or enhance existing frameworks.

Canada also has a strong privacy framework through the Personal Information Protection and Electronic Documents Act, PIPEDA, at the federal level, alongside additional provincial laws. The Canadian law sets rules for businesses regarding the collection, use, and disclosure of personal information in the context of commercial activity. Canada has also been acknowledged by the European Union as offering a sufficient level of protection, facilitating fairly seamless data transfers between Europe and Canada.

The Canadian government is in the process of revising its legislation through a new bill, formerly referred to as Bill C-11, to enhance individual rights in the digital age. This involves giving individuals the right to request the deletion of their data and updating the understanding of consent to align with contemporary developments.

In East Asia, South Korea stands out as a country that takes data privacy very seriously. It has a strong law known as the Personal Information Protection Act, PIPA, enacted in 2011 and amended several times to strengthen its requirements. South Korea established the Personal Information Protection Commission (PIPC) as an independent regulatory authority. This commission has demonstrated strict enforcement, even against global technology giants.

In 2022, South Korea imposed huge fines on both Google and Facebook/ Meta for privacy violations, involving the collection and analysis of users' behavioral data for advertising purposes without clear consent. The fines reached about 69.2 billion won for Google, around 50 million dollars, and 30.8 billion won for Meta, around 22 million dollars. These figures reflect the seriousness of Korean authorities in protecting citizens' privacy and penalizing violations regardless of the size of the company. Many observers praised these actions and viewed them as a model that other countries can follow to regulate the practices of large corporations. Turning to major data breach incidents, they have varied over the past years in their nature and scale, but they have all sounded the alarm about the importance of information privacy and security.

One of the most well-known privacy scandals in the digital era is the Cambridge Analytica and Facebook scandal in 2018. In this incident, it was revealed that a British political consulting firm called Cambridge Analytica obtained the personal data of about 87 million Facebook users without their knowledge, through a personality quiz application that was built on Facebook's platform. That data was then used improperly to influence voters' opinions in major political and electoral campaigns, such as the United States elections and the referendum on the United Kingdom leaving the European Union. The scandal exposed how the exploitation of social media data can pose a threat even to democracy itself. It led to investigations and financial penalties, including legal settlements that reached 725 million dollars against Facebook, now Meta. Most importantly, it pushed platforms to reassess their policies on granting third-party applications access to users' data.

Another striking example is the Careem breach in 2018 in the Middle East. Careem is a popular ride-hailing service similar to Uber, operating in the Gulf and other regions. It was hit by a cyberattack in which attackers managed to break into its systems and access a large database containing information on around 14 million users and drivers. The leaked information included names, email addresses, and trip history. However, the company stated that passwords and credit card data were not stolen, as payment information was stored on external systems. Even so, the breach raised widespread concern among users in the region due to the scale of the incident, and it pushed regional companies to increase investment in cybersecurity and privacy protection. Careem apologized to its users, emphasized that it had learned from the incident and strengthened its systems, while many customers changed their passwords and monitored their accounts out of fear of potential misuse. The case was a reminder that local technology companies are not immune to attacks, and that Arab users' data has real value and can be a target for hackers worldwide.

No discussion of data leaks is complete without mentioning the incident that is still considered the largest in history in terms of the number of affected accounts, the hack of the well-known Yahoo service. In 2013, Yahoo suffered a catastrophic breach in which hackers stole data for nearly all user accounts on the platform, estimated at 3 billion accounts worldwide. The stolen data included usernames, email addresses, encrypted passwords with weak encryption that could be cracked, and other personal information. What was most shocking was that Yahoo did not fully understand the scope of the breach until years later. At first, it announced that one billion accounts were affected, then in 2017, it acknowledged that the true number was all existing accounts. This incident caused a major drop in Yahoo's market value and forced the company to be sold at a lower price. Yahoo also faced dozens of lawsuits and regulatory penalties. The lessons were harsh; even internet giants are not immune, and negligence in updating security systems, or weak encryption, can lead to massive leaks that may remain undiscovered until it is too late.

In addition to the aforementioned examples, numerous other significant incidents have occurred, including the 2017 Equifax breach, which compromised sensitive financial information for approximately 147 million individuals, and the extensive data breach at Marriott, discovered in 2018, which affected hundreds of millions of hotel guests and included passport information. More recently, there have been alarming incidents involving major social media platforms, such as the 2020 hacking of multiple verified Twitter accounts to facilitate a cryptocurrency scam, among others. These events illustrate that no entity is entirely immune to the risk of a data breach. Consequently, institutions must invest substantial resources in securing their systems, while individuals should remain vigilant, activate security features available for their accounts, and monitor alerts and news related to potential breaches involving services they utilize, enabling them to take immediate actions such as altering passwords or scrutinizing their financial records.

In response to the increasing prevalence of privacy infringements and data breaches, there has been a global movement towards the establishment of more stringent regulations and elevated standards for organizations. It is now deemed unacceptable for entities to collect personal data without a clear framework of accountability, or to neglect the safeguarding of such data. Although privacy laws may vary from one jurisdiction to another, they are unified by a central objective: to uphold individuals' rights to privacy and control over their personal information, while mandating organizations to protect that information from misuse or unauthorized access.

As these regulations have evolved, companies have also begun to adopt more transparent practices with users. Today, it is common to encounter privacy setting prompts on websites and applications, along with options to accept or decline cookies. This development represents a positive progression towards empowering individuals within the digital landscape.

Children's and Teenagers' Online Privacy: What Parents Need to Know

In today's environment, where children are increasingly immersed in technology from a young age, the issue of digital privacy for younger generations has emerged as a critical concern. Children and adolescents engage with the internet for various purposes, including gaming, educational pursuits, social interactions, and school-related activities, exposing them to many of the same privacy challenges that adults encounter, often to a greater extent. A noteworthy statistic from the European Commission indicates that one in every three internet users worldwide is a minor under the age of eighteen. This underscores that a significant portion of the digital landscape is inhabited by young users who may lack a comprehensive understanding of the associated risks or the necessary experience to safeguard themselves. The trend of children accessing the internet at increasingly younger ages, through a myriad of devices such as tablets, smartphones, and computers, often results in prolonged connectivity, frequently occurring without direct parental oversight. Consequently, educating both children and their parents regarding digital privacy is no longer a luxury but an urgent necessity.

The implications of digital privacy for children and teenagers can be viewed from two distinct perspectives. On one hand, children are particularly susceptible to violations and exploitation, often sharing personal information naively, without awareness of the potential ramifications. On the other hand, infringements upon their privacy can lead to enduring psychological and social consequences.

Several examples of inherent risks are as follows:

1.   Exposure to Inappropriate Content and Cyberbullying:

Children may inadvertently encounter violent or sexually explicit material, as well as malicious messages that can cause emotional harm. Additionally, they may fall victim to cyberbullying, which can manifest as harassment and abusive comments on social media platforms or during online gaming sessions. Such bullying may involve the unauthorized disclosure of private information or the unconsented sharing of photos for ridicule, undermining the child's privacy and dignity.

2.   Online Predators and Manipulators:

The internet, unfortunately, provides abusers and fraudsters with novel avenues to target vulnerable youth. An individual may impersonate a child or teenager, engaging with one's son or daughter via chat applications or gaming platforms, gradually fostering trust over time, before soliciting sensitive information such as home addresses, phone numbers, or details about their educational institutions. This information can subsequently be exploited to endanger the child or to facilitate in-person meetings arranged without parental knowledge. Numerous cases have been documented where offenders successfully made contact with minors by exploiting acquired information.

3. Loss of Control Over Personal Data at an Early Age:

As adolescents establish social media accounts and publicly share their thoughts and images, they unintentionally create a digital footprint that may persist well into adulthood. An ill-considered post or a misjudgment made during this developmental phase can resurface later, potentially impacting opportunities for university admissions or employment, as online content is frequently archived unless actively deleted. Moreover, companies may collect significant quantities of data concerning minors for marketing endeavors without their awareness, utilizing targeted advertising based on viewing habits or digital games that gather behavioral and locational information for distribution or sale. This poses a challenge, as children often lack the discernment to differentiate between advertising and authentic content, and may not fully grasp the implications of data utilization.

4. Lack of Knowledge or Ability to Navigate Privacy Settings:

Many applications and websites are primarily designed with adults in mind; consequently, privacy settings may be overly intricate for younger users. A child may be unaware of how to make their accounts private or how to block individuals who are harassing them. Even if they possess the knowledge, they may refrain from informing their parents out of fear of repercussions or a loss of access to their devices. As a result, they may endure harassment, violations of privacy, or predatory grooming without seeking assistance.

Given these substantial risks, parents and educators play a pivotal role in safeguarding children's online privacy. The following guidelines should be well understood and implemented by caregivers and educators to enhance the safety of young users within the digital domain:

1.   Promote Awareness and Open Communication:

Facilitate regular discussions about internet etiquette, its associated risks, and the significance of privacy as a routine part of conversations with children. Emphasize in straightforward terms that, similar to how they avoid sharing personal information in certain physical spaces, they must also refrain from disclosing such details online to strangers. Encourage children to report any unsettling experiences or content encountered online without fear of punishment, assuring them that you are a supportive confidant in times of distress. Reinforce their understanding that not all online information or individuals are trustworthy.

2. Establish Clear Guidelines for Usage:

Before granting children access to devices or permitting registrations on websites or applications, it is essential to establish explicit family rules regarding safe usage. For instance, mutually agree on specific time limits for device use.

3. Creating an Environment of Understanding:

By fostering an environment of understanding and diligence, caregivers can bolster the protective measures surrounding digital privacy for children and teenagers.

For instance, you can configure your home Wi Fi so your child's devices lose access after 9 pm to support healthy sleep routines. There are also specialized monitoring programs that alert you if a child visits suspicious website or attempts to share personal information. Some tools provide weekly reports showing how the device is being used. However, parents should use these tools wisely and transparently. Tell your child you are using them, and explain why, to ensure their safety, not to spy on them. This helps the child understand that it is about protection, not a lack of trust.

4. Teach Responsible Online Behavior:

Just as we teach children manners and respectful behavior in real life, we must teach them digital etiquette as well. Instill the principle of respecting another people's privacy, too. Children should never attempt to hack a classmate's account or share someone else's information without permission. Teach them that if they would not

accept something happening to them, they should not do it to others online. Encourage them to think before posting: would they still be proud of what they wrote a year from now? Could it hurt someone? Also, warn them not to respond to harassment or hostile messages. Instead, they should immediately inform a trusted adult, whether a parent or a teacher, so appropriate action can be taken, such as blocking or reporting the bully. When children see that you take their concerns seriously, they will be much more willing to come to you in the future.

5.   Protection of Children's Privacy:

Protect children's privacy even before they can protect it themselves: This advice is especially important for parents. As caregivers, we also have a responsibility to respect our children's privacy in the digital world. For example, the growing practice of "sharenting," where parents frequently post their children's photos online without considering the child's future wishes, can become a privacy violation in itself. A child may grow up to find that many details of their childhood have been publicly shared on the internet. For this reason, it is best to limit posting children's photos and personal information on public platforms as much as possible, and to share only what is necessary or harmless within private, closed circles of family and trusted friends. Remember that any image you share of your child could reach people with harmful intentions and be misused in inappropriate ways.

Also, make sure that any app you give your child, such as mobile games, is age-appropriate, comes from a reliable source, and has been reviewed carefully. Read user reviews and check the developer's credibility, since some apps have been found to track children or collect their data without clear permission. In short, make protecting your child's digital privacy part of your parental responsibility, just as you protect them in every other area of life.

It is also important to note that there are laws and regulatory considerations designed specifically to protect children's privacy. For example, in the United States, the Children's Online Privacy Protection Act, known as COPPA, restricts the collection of personal data from children under 13 without parental consent. This law requires child-directed websites and apps to obtain explicit permission from parents before collecting any identifying information, such as a name or address. It also requires clear, simplified privacy policies and gives parents the right to review the data or request its deletion. Many global platforms, even outside the United States, effectively follow the spirit of this law to protect their reputation. This is why many major social media platforms officially prohibit users under 13 from registering. Although some children bypass these rules with adult help, such laws still serve as important safeguards. In Europe, the GDPR also includes specific provisions related to children, recognizing that minors need additional protection when it comes to personal data.

Ultimately, protecting children's privacy online requires a combined effort of legislation, education, and ongoing family guidance. The parents' role is not limited to monitoring and restricting; it is also about building trust and shared understanding. When a child feels safe talking about what they experience online, and when they develop critical thinking and personal caution, we have truly equipped them to become responsible and aware users in a fast-moving digital world. And we should remember that children are often more capable than we expect when it comes to learning technology. With the right guidance, they may surprise us with how well they can judge situations and act wisely, even in the vast space of the internet.

The Importance of Digital Awareness for Protecting Privacy in a Rapidly Evolving Digital Environment

As we conclude this chapter, a pivotal message emerges: digital awareness is the foundational line of defense in safeguarding privacy within our interconnected landscape. We have delved into the significance of digital privacy as a fundamental human right and an integral component of digital wellness. Our exploration has covered the vastness and variety of personal data, categories of data collection, and the multifaceted applications of this data, elucidating both the beneficial outcomes and the potential risks we face. Additionally, we examined various tools and practices that empower individuals, regardless of age or technical expertise, to take proactive measures in securing their privacy.

The digital realm is advancing at an unprecedented rate, with new technologies and applications surfacing daily. This rapid evolution introduces new privacy challenges. We often feel an erosion of control due to ubiquitous surveillance, ubiquitous data-hungry applications, and relentless data requests from platforms. However, it is crucial to recognize that maintaining control over personal privacy remains achievable by leveraging knowledge and applying available tools judiciously.

Awareness of the data we share and the entities with which we share it, along with a cautious approach to adopting new services and critically evaluating permission requests, should be treated with the same seriousness as maintaining physical and mental health in an overstimulating environment. Protecting one's privacy in a data-saturated digital landscape is as vital as safeguarding personal wellness.

We also discussed the encouraging trend of an increasing number of legal and regulatory frameworks globally aimed at bolstering privacy protections. This signifies a growing acknowledgment of user concerns and a concerted effort to rebalance the power dynamic between large tech corporations and individual rights. Nonetheless, reliance on legislation alone is insufficient. Enforcement mechanisms can be slow, and privacy breaches may occur even with comprehensive regulations in place. Hence, individual and collective accountability remain paramount. Individuals should pursue education about privacy, organizations must instill a culture of privacy by design rather than merely compliance-driven practices, and governments need to adapt legislation flexibly to keep pace with evolving technologies.

It is essential to recognize that digital privacy is neither an infrequent luxury nor an obstacle to innovation. Instead, it serves as the bedrock of trust necessary for a thriving digital ecosystem. Without assured data security, confidence in digital services diminishes, curtailing their potential utility. Conversely, when users believe their data is handled responsibly, they are more likely to engage positively and proactively with technology. Achieving this equilibrium necessitates collaborative efforts across technical, legal, and educational domains. Every stakeholder has a role—be it parents educating their children, educators embedding privacy concepts into curricula, informed youth guiding peers in privacy settings, or consumers advocating for more ethical data practices from corporations.

In summary, safeguarding your digital privacy today equates to preserving a facet of your freedom and personal security for the present and future. Remain vigilant, apply the principles and tools you have learned, disseminate this knowledge within your networks, and continue to adapt as technology evolves. Digital privacy is both a right and a responsibility. As awareness increases and responsible practices proliferate, we collectively contribute to cultivating a digital environment that harmonizes technological advancement with the protection of human dignity and fundamental rights. Your privacy constitutes your personal realm—value it, as it is deserving of rigorous protection.

Chapter Six:

"Children and Digital Wellness"

Introduction to the Importance of Digital Wellness for Children

With rapid digital transformation reshaping almost every aspect of modern life, paying close attention to children's digital wellness has become essential and inseparable from their upbringing and overall health. Today's children grow up surrounded by smart devices, internet connectivity, and digital platforms, often beginning their digital engagement at very early ages. Global reports indicate that nearly one third of internet users worldwide are children and adolescents, a figure that highlights the scale of young people's presence in digital spaces. In the United Arab Emirates, for example, recent data suggests that approximately 97 percent of children aged seven and above use digital devices regularly. While this widespread access to technology offers enormous benefits for education, creativity, and communication, it also introduces challenges and risks that require conscious awareness and thoughtful guidance.

Parenting in the digital age has become significantly more complex due to the speed of technological change and the pressures created by the digital revolution. Parents are no longer only responsible for guiding their children's behavior in physical environments, but must also navigate virtual spaces that are constantly evolving. Achieving a healthy balance between benefiting from technology and protecting children from its harmful effects lies at the heart of digital wellness. Just as parents actively safeguard a child's physical and mental health, they must also protect their digital health and guide their online behavior. Digital wellness recognizes that children's experiences online can shape their emotions, habits, values, and social relationships in powerful and lasting ways.

The importance of digital wellness for children stems from the fact that technology has become deeply embedded in daily life, influencing educational experiences, social interactions, and psychological development. Without proper guidance and boundaries, children may be exposed to inappropriate content, cyberbullying, online exploitation, or harmful social comparisons, all of which can leave lasting negative effects on behavior and emotional wellness. International organizations have repeatedly warned of these risks. UNICEF, for instance, has emphasized that while digital access provides children with valuable opportunities for learning and participation, it also exposes them to serious dangers, including cyberbullying and various forms of online violence that can occur at any moment in the digital environment.

At the same time, when technology is introduced in a safe, structured, and supportive manner, it can open meaningful opportunities for learning, creativity, and positive social engagement. Children can develop digital skills, explore knowledge beyond their immediate surroundings, and build confidence in using modern tools responsibly. For this reason, digital wellness is not about restricting technology altogether, but about ensuring that children grow up with the awareness and self-control needed to use it wisely. It is a fundamental component of raising a balanced generation that is capable of benefiting from technology without becoming overwhelmed or harmed by it.

This chapter presents a comprehensive discussion of healthy digital parenting. It explores children's digital needs across different age groups and outlines practical household rules for managing technology use, supported by examples of age-appropriate applications and games. It also examines the effects of unguided and excessive digital exposure, drawing on real stories from the Gulf region and the wider world, including both success stories and cautionary examples that offer valuable lessons. In addition, the chapter discusses how parents can teach children to protect themselves online with awareness and growing independence. It reviews key Gulf and international initiatives aimed at safeguarding children on the internet and highlights the important role played by schools and teachers, incorporating insights from specialists in education and technology. Finally, it addresses common parenting mistakes related to children's technology use. The overarching goal is to provide a practical, awareness-based guide that strengthens our understanding, as parents, educators, and young adults, of the importance of digital wellness in everyday life and equips us with the tools needed to raise a digitally aware, confident, and safe generation.

The Concept of Healthy Digital Parenting and Its Pillars

Healthy digital parenting refers to equipping children with the knowledge, values, and skills that enable them to engage with technology in a safe, ethical, and balanced manner. Many experts define digital parenting as a set of habits, values, skills, and codes of conduct that regulate engagement with digital and virtual technologies. In this sense, it functions as a framework for behavior in the digital world, much like social upbringing organizes behavior in the physical world. When this form of guidance is absent, digital spaces can become chaotic environments in which harmful behaviors spread easily. Over time, this lack of structure can contribute to digital addiction and negatively affect an individual's academic performance, social relationships, and daily functioning. For this reason, healthy digital parenting has become an urgent necessity in an era of rapid technological advancement.

The concept of digital parenting can be understood through several core pillars that together form a comprehensive framework for children's digital development.

1. Digital Safety and Security:

This pillar focuses on protecting children from online risks such as inappropriate content, cyberbullying, fraud, and exploitation. It includes teaching children basic principles of online safety, such as not sharing personal information with strangers, using privacy settings effectively, and informing trusted adults when they encounter harassment or uncomfortable situations online. Parents are often advised to strengthen privacy settings to limit contact with unknown users, monitor screen time and content appropriately, review a child's online activity, and teach children how to recognize suspicious links or unfamiliar contacts. This pillar also includes legal and rights-based awareness, such as helping children understand their digital rights, including the right to privacy, as well as their responsibilities when interacting online.

2.   Media Literacy and Digital Awareness:

This pillar emphasizes developing strong digital literacy skills so that children can distinguish between accurate and misleading information and identify trustworthy sources. It encourages critical thinking about online content and teaches children how to question what they see, read, and hear on digital platforms. Media literacy also includes practical skills, such as understanding how personal information shared online contributes to a lasting digital footprint. Healthy digital parenting keeps pace with technological change by ensuring that children learn responsible digital values before they engage deeply with devices and social platforms. This early awareness helps them navigate digital spaces with confidence rather than confusion.

3.   Digital Behavior and Ethical Responsibility:

While technology provides access to information and communication, it also introduces new social norms and ethical challenges. Teaching children respectful and responsible digital behavior is therefore essential. This includes promoting kindness, respectful communication, and empathy in online interactions, as well as discouraging harmful behaviors such as cyberbullying, hate speech, or harassment. This pillar also addresses online etiquette, respect for intellectual property, and awareness of the impact one's actions can have on others. Research consistently shows that digital parenting should foster values that guide how technology is used, allowing individuals to benefit from digital tools without harming real-life relationships or personal wellness. Ultimately, the goal is to raise responsible digital citizens who respect themselves and others in both online and offline spaces.

4. Digital Balance and Mental Wellness:

Maintaining a healthy balance between digital life and real life is one of the most critical pillars of digital wellness. This includes managing screen time so that it does not interfere with sleep, academic responsibilities, physical activity, or family interaction. Although digital devices are deeply embedded in children's daily routines, their use must remain balanced and purposeful. Healthy digital parenting establishes clear limits and encourages offline activities such as sports, creative hobbies, and social play. International health organizations emphasize this balance. The World Health Organization, for example, recommends no screen time for children under the age of one and no more than one hour per day of high-quality content for children aged two to five. These guidelines aim to protect children's physical and mental development during critical early years, when sleep, movement, and human interaction are especially important.

5. Communication and Trust Between Parents and Children:

Digital parenting cannot succeed without open communication and mutual trust. Children should feel safe discussing their digital experiences, questions, and concerns without fear of punishment or judgment. This pillar encourages parents and educators to remain actively involved in children's digital lives while offering consistent guidance and emotional support. Research and experience suggest that children are more resilient when they grow up in environments where open dialogue is encouraged. Regular communication allows adults to address issues early, correct misunderstandings, and provide timely advice. It also empowers children with the confidence to seek help when facing online challenges.

When these pillars work together—safety, awareness, ethical behavior, balance, and communication, they form a comprehensive and effective approach to digital parenting. Healthy digital parenting is not limited to technical rules or restrictions; it is an ongoing process of instilling positive digital values and habits from an early age. Just as children learn boundaries and social norms in the real world, they must also learn boundaries in the digital world so they can benefit from technology without allowing it to dominate or harm their lives. As the saying goes, digital parenting in the virtual world is equivalent to social upbringing in the physical world. It is the foundation for raising a generation capable of navigating digital spaces with confidence, responsibility, and care.

Age Groups and Their Different Digital Needs

Children's digital needs vary significantly depending on their age, developmental stage, and level of maturity. Understanding what is appropriate for each stage allows parents and educators to provide guidance that supports healthy growth rather than hindering it. Below is a general overview of key age groups and their specific needs in the digital environment.

1. Early Childhood and Preschool Age (Approximately 0 to 5 Years)

During early childhood, children's understanding of technology is very limited. Their attention is primarily drawn to bright visuals, sounds, and simple interactive elements. Experts consistently recommend that screen exposure for infants and toddlers should be minimal or avoided altogether. According to World Health Organization guidelines, children under the age of one should not be exposed to screens, while children aged two to five should have no more than one hour per day of high-quality digital content. At this stage, children rely heavily on direct human interaction, physical play, and consistent sleep patterns to support healthy brain development.

Digital Needs at This Stage: At this age, any digital exposure should be purposeful, limited, and carefully supervised. Content should be simple, educational, and age appropriate, focusing on basic concepts such as letters, numbers, shapes, and sounds. Applications like Endless Alphabet or Arabic alphabet learning apps can provide engaging introductions to language skills. Platforms such as Khan Academy Kids offer structured educational content designed specifically for children under five, combining stories, games, and early learning activities.

Equally important is co-use with parents. Children at this stage benefit most when a parent is actively involved in their digital experience. Watching educational videos together, playing simple interactive games, or reading illustrated digital storybooks as a shared activity helps transform screen time into a social and educational experience rather than a passive one. The American Academy of Pediatrics strongly recommends shared viewing during early childhood, emphasizing that interaction and discussion enhance understanding and reduce the isolating effects of screen use.

Content Safety and Limited Advertising: Parents should ensure that the applications and videos used by young children are free from advertisements and in-app purchases, as children at this age cannot distinguish between content and commercial material. Parental controls should always be activated to prevent exposure to inappropriate content, including within video platforms. Many families prefer to use platforms such as YouTube Kids, which are designed to offer a safer and more age-appropriate viewing experience. However, it is important to remember that even content labeled as "child-friendly" requires occasional monitoring to ensure quality, accuracy, and suitability.

At this early stage, screen time should remain very limited. The primary focus must be on developing motor skills, language abilities, and social interaction through real-world play and direct engagement with people. At best, technology should serve as a simple supporting tool for learning and limited entertainment, not as the main source of education or stimulation. Establishing healthy habits from the beginning is also essential. Parents are encouraged to avoid device use during meals or before bedtime, as such habits become increasingly difficult to change later in life. Devices should never be used as tools to stop crying or calm tantrums, as this may create an unhealthy association between screens and emotional regulation. Instead, children should learn to manage emotions through human interaction, comfort, and guidance.

2. Primary School Age (6 to 12 Years):

As children enter primary school and move into middle childhood, their curiosity expands and their cognitive abilities develop rapidly. At this age, children acquire foundational skills in reading, writing, and basic mathematics, and technology can become a valuable educational support when used appropriately. Many children in this age group have access to tablets or use the family computer under parental supervision. Some also begin watching cartoons online or playing age-appropriate video games.

Digital Needs at This Stage:

- *Learning and Educational Games:*

Children at this age can benefit greatly from structured educational applications that present learning in engaging and interactive ways. Examples include applications that teach mathematics, science, and problem-solving skills through games and challenges. The well-known Khan Academy platform is suitable for various age groups and offers interactive lessons in mathematics, science, and other subjects aligned with school curricula. Applications such as Epic provide large digital libraries of children's books, encouraging reading and improving language skills. To strengthen logical thinking, puzzle and reasoning games such as Busy Shapes can support visual perception and problem-solving development. For language learning, simplified tools such as Duolingo for Kids introduce new vocabulary in a playful, game-like format that sustains motivation.

- ***Developing Creativity and Basic Technical Skills:***

Children in this age group often display strong imagination and a desire to create rather than simply consume content. Parents and educators can guide them toward digital tools that nurture creativity, such as drawing, coloring, and story-creation applications. This stage is also ideal for introducing basic coding concepts through simple, interactive games. For example, Hopscotch allows beginners to learn how to build games and interactive stories in an enjoyable way. Visual programming platforms such as Scratch, developed by MIT, provide a safe and child-friendly environment where children can create projects and share them within a moderated community. These activities help shift children from passive content consumption to active creation, strengthening confidence, problem-solving skills, and a sense of achievement.

- ### *Age-appropriate Entertainment Content:*

It is natural and acceptable for children to enjoy digital entertainment when it is appropriate and well chosen. Parents should carefully select cartoons and programs that are suitable for the child's age and values. Video games should always match the recommended age ratings, and parents can rely on established rating systems such as PEGI or ESRB to guide their choices. Violent games or content designed for older age groups should be avoided. Some games can be beneficial when used with proper settings and time limits. For example, Minecraft is a building and exploration game that supports creativity, cooperation, and problem-solving, and has been widely used in educational contexts. Light puzzle games, adventure games, or mathematics-based games can successfully combine enjoyment with learning.

- ### *Ongoing Supervision and Guidance:*

Although children in this age group are more independent than preschoolers, adult supervision remains essential. Clear household rules help create structure and predictability. For example, parents may allow device use for a set daily period, such as one to two hours, depending on the child's age, school workload, and responsibilities. Another common rule is that internet use should occur only in shared family spaces, or that parental approval is required before downloading any new application or game. Keeping devices in common areas rather than in closed bedrooms allows parents to remain aware of online activity without constant monitoring. Internet filters, safe search features, and parental control tools should be enabled to block inappropriate content and restrict access to unsuitable sites or applications.

At this stage, children are enthusiastic about technology, and their curiosity may lead them to experiment without fully understanding consequences. Therefore, guidance should focus on continuous awareness delivered in simple, age-appropriate language. Adults can explain why personal information should not be shared online and why screen time is limited rather than unlimited. At the same time, it is important to strengthen offline activities such as sports, social play, reading, and creative hobbies, ensuring that digital devices do not become the sole source of enjoyment. Building this balance during middle childhood lays the foundation for healthy and responsible technology use later in life.

3. Adolescence (13 to 18 Years)

As children enter adolescence, their relationship with technology becomes more complex and deeply intertwined with identity, independence, and social interaction. Teenagers often receive their first personal smartphone during this stage and become more actively engaged in social media and online communication with peers. Adolescents today play a major role in global digital connectivity. United Nations statistics indicate that approximately 79 percent of individuals aged fifteen to twenty-four use the internet regularly. This widespread engagement makes it especially important to understand and address adolescents' specific digital needs.

Digital Needs at This Stage:

- ### *Independence within a Framework of Trust:*

Adolescents naturally seek autonomy and want greater control over their decisions, including how they use technology. Healthy digital parenting at this stage requires granting trust and independence while maintaining agreed-upon boundaries that ensure safety. Rather than imposing rigid rules, guidelines can be discussed and negotiated. Specialists emphasize that digital boundaries should evolve as children grow, allowing teenagers to gain independence gradually within safe limits. For example, a teenager may be allowed additional screen time on weekends or permitted to use a computer in their bedroom for schoolwork, provided they demonstrate responsible behavior and communicate openly about online challenges.

- ### *Safe Digital-Social Interaction:*

Social media platforms such as Instagram, Snapchat, TikTok, and others play a central role in teenagers' social lives. This makes it essential to teach responsible communication and digital privacy protection. Age requirements should be respected, as most platforms require users to be at least thirteen years old. Teenagers should understand that public accounts make content visible to a wide audience, and that setting accounts to private or limiting visibility to trusted contacts is safer. They should be advised not to accept friend requests from strangers and to avoid sharing sensitive personal information or private images. If suspicious messages, fake job offers, or unknown links appear, teenagers should be encouraged to report and block the sender immediately and inform a trusted adult. Many teenagers already practice this awareness. For example, a thirteen-year-old student reported receiving fake discount links and false job offers on WhatsApp and stated that she immediately blocks and reports them.

- ***Awareness of Behavioral and Psychological Risks:***

Adolescence is marked by heightened sensitivity to peer opinion and social comparison. Teenagers may be particularly vulnerable to cyberbullying, online pressure, and unrealistic comparisons promoted by social media. Global studies suggest that more than one third of young people across multiple countries have experienced cyberbullying, and that a significant number have avoided school as a result. Teenagers need resilience and self-confidence to cope with such challenges. They should be reminded that personal worth is not measured by likes or online approval, and that abusive behavior reflects the actions of the aggressor rather than the value of the victim. Teenagers should also be made aware of harmful online content that promotes violence, despair, or self-harm, and they should feel encouraged to seek support from parents, teachers, or counselors when digital experiences negatively affect their emotions.

- ***Constructive Use and Productivity:***

Adolescents possess energy, creativity, and the ability to use technology for meaningful purposes. Parents and educators can guide them toward productive digital engagement, such as online courses in areas of interest like design, coding, photography, or languages. Interactive learning platforms can also support academic achievement. Teenagers may be encouraged to undertake constructive projects, such as creating a personal blog, educational content, or a purposeful social media channel, while receiving guidance on ethical content creation and safe interaction. In the Gulf region, for example, some teenagers have developed applications that serve community needs or launched online awareness campaigns, gaining recognition for their contributions. These examples demonstrate that the digital world can be a space for creativity, innovation, and social impact, not merely entertainment.

- ***Digital Privacy and Legal Responsibility:***

Teenagers must understand that they are responsible for what they publish and share online, and that digital actions can carry ethical and legal consequences. Sharing private images or personal data without consent can result in serious harm and legal accountability. Clear boundaries should be established regarding the use of cameras and recording tools, emphasizing respect for others' privacy and dignity. Introducing adolescents to local cybercrime and information security laws can reinforce this awareness. In the United Arab Emirates, for instance, Wadeema's Law criminalizes the online exploitation of children and obligates service providers to report harmful content. Awareness of such laws can deter harmful behavior and protect teenagers from exploitation.

In general, supporting adolescents in the digital world requires a careful and thoughtful balance. Adults must offer trust, guidance, and emotional support while remaining attentive to the challenges teenagers may encounter online. The most effective approach is to build a relationship based on openness and ongoing dialogue about technology. When teenagers feel safe discussing their digital experiences without fear of excessive punishment, they are more likely to accept guidance and follow safety practices. Although adolescents may appear highly skilled with devices, they still rely on adult experience and judgment to help them make responsible and informed choices in an increasingly complex digital environment.

Effective Household Rules for Managing Children's Use of Technology

Setting clear rules for technology use at home is the foundation for supporting children's digital wellness. These rules help children understand boundaries and develop self-discipline in how they use devices. Below are practical guidelines and recommendations that have proven effective for families in managing children's technology use.

1.　Set Time Limits and Use Screen Timers:

It is essential to establish a daily or weekly time limit for using electronic devices, such as television, tablets, smartphones, and video games. The appropriate duration varies by age. For example, one hour per day may be suitable for younger children, while teenagers may be allowed more time, with additional flexibility on weekends. The goal is to prevent excessive use that can contribute to digital addiction or interfere with other priorities, such as sleep, study, and physical activity. Some guidance suggests that entertainment screen time should not exceed two hours per day for children, with limited increases for older children, as long as responsibilities are met. Families can utilize tools like timers or screen time management applications to promote consistency.

2.   Create Technology-Free Times and Spaces:

Families benefit from shared time that is not interrupted by screens. For example, agree that the dining table is completely free of phones and television so that meals become a time for family conversation. You can also set one evening each week as a family night without electronics, dedicated to board games or an outdoor activity. These digital breaks teach children the value of face-to-face connection and help create positive family memories without screen interference. Specialists also recommend a no devices in the bedroom rule, especially during the hour before sleep, because blue light can reduce sleep quality. For this reason, it is often better to charge phones and devices outside children's bedrooms at night.

3.   Be a Positive Role Model as a Parent:

Children learn habits by observing their parents. If adults set rules for children, they should follow the same rules themselves. It is inconsistent to ask a child to put a phone away during dinner while a parent is using a phone at the table. Try to reduce device use in front of children and demonstrate positive uses of technology, such as reading, learning, or completing tasks, rather than spending all free time on social media. When children see parents balancing digital life with real life, they are more likely to learn and adopt that essential skill.

4. Agree on Rules Together and Define the Consequences of Breaking Them:

Children should be involved in setting the rules for device use. A simple family meeting where parents explain their concerns and listen to the child's preferences can lead to a clear agreement. For example, you might agree that playing on a console is allowed for one hour after homework, or that internet use is limited to the living room under adult supervision. When children participate in the process, they feel a greater sense of responsibility and are more likely to follow the agreement because they helped shape it. It is also helpful to write the rules clearly and place them somewhere visible at home. Some families use a Family Internet Agreement. This agreement should also specify the consequences of breaking the rules so they are known in advance. Consequences might include losing device access for a day or reducing certain privileges. The key is that consequences are reasonable and consistent, so children take them seriously. It is also recommended that all adults in the home apply the same policy, so children do not receive confusing exceptions from one person but not another.

5. Set Clear Locations for Use and Monitor Appropriately:

Designate specific places where children may use devices. For example, a desktop computer can be kept in the living room or another open space, rather than in a bedroom where the child is completely alone. For mobile devices, families can set a rule that they should not be used behind closed doors. The goal is to allow occasional visual monitoring without violating the child's sense of privacy. For younger children, devices should also be kept out of reach except during allowed times. For example, a parent can keep the tablet and provide it only during approved use periods. These simple steps reduce the likelihood of misuse.

6.   Use Parental Controls Wisely:

Many devices and applications now offer parental controls that allow families to filter content, set time limits, and monitor activity. It is useful to benefit from these tools, such as adjusting device settings to block inappropriate content or using dedicated tools like Google Family Link or Norton Family to support online safety and screen time management. Specialists advise explaining to older children, clearly and respectfully, why these tools are being used, so they do not interpret them as a lack of trust. Tell your child that the purpose is protection, not surveillance. For example, Norton Family describes several key benefits of parental controls, including cybersecurity, time management, protecting digital reputation, data backup, and online etiquette. Explaining these ideas to older children can help them see that parental controls are a safety measure meant to support them, rather than an invasion of digital privacy.

7. Promoting Positive Use Rather Than Absolute Bans:

Experts in digital parenting emphasize adopting a balanced perspective toward technology. Two approaches are often unhelpful. The first is treating technology as entirely harmful and banning it completely. The second is ignoring the risks and allowing unrestricted use. Neither approach serves the child well. The goal is balance. Children should learn that technology, like any tool, can be used in constructive ways or in harmful ways. Instead of relying on prohibition alone, the focus should be on guidance. Yes to a beneficial game, but no to excessive use. Yes to a cultural or educational program, but under supervision. This approach helps children develop personal awareness of boundaries rather than feeling that limits are imposed by force.

8. Strengthening Dialogue and Ongoing Awareness:

Make conversations about technology part of everyday family discussion. Ask your children questions such as, what game are you playing these days, what videos have you watched recently, and who are your online friends. Share your own experiences simply and appropriately as well. When dialogue remains open, children feel comfortable telling you about anything upsetting they encounter online instead of hiding it out of fear of punishment. For example, if a child receives a strange message or sees inappropriate content, encourage them to come to you immediately. Emphasize that you will support them and that you will handle the situation together, whatever happens. This trust is a true safety safeguard. When families apply these rules and guidelines, the home becomes a more organized and healthier digital environment. Children feel safe and clear about what is allowed and what is not, which helps them develop positive digital habits that can last a lifetime. A familiar saying captures this idea: a child grows up shaped by what they become used to. When children grow up with digital self-discipline from an early age, they are more likely to manage their digital lives independently and responsibly as they mature.

Examples of Safe and Appropriate Applications and Games for Children of Different Ages

The digital marketplace offers many applications and games designed for children. The key is to choose what matches the child's age and interests and to ensure it is safe and educational. Below are suggested examples by age group, with the reminder that parents play an essential role in testing and evaluating what children use.

Preschool Age, Educational Applications, and Interactive Stories

Khan Academy Kids: A free, comprehensive educational application designed for children up to age five. It offers interactive content in mathematics, reading, and science through stories, games, and songs. It is currently available in English and uses an engaging approach that supports early cognitive and language development.

Abjad, for learning Arabic letters: An Arabic language application for teaching letters and spelling, suitable for ages two to five. It uses attractive visuals and interactive activities that capture children's attention and teach basic Arabic language skills in a safe environment that is described as free of advertising and inappropriate content. Endless Alphabet: An English language application that teaches letters, sounds, and simple vocabulary through playful cartoon characters. It supports letter recognition and pronunciation through word-building games.

Lamsa: A comprehensive Arabic application that offers a library of stories and educational games for young children. It includes a large collection of interactive activities, such as stories, songs, and learning games, presented in a generally child-friendly environment with content oversight.

YouTube Kids: A video platform designed specifically for children, with a simplified interface and content presented as age-appropriate. Children can watch their favorite programs and educational or entertainment videos suited to their age. The platform also includes parental controls that allow parents to limit what can be watched and how long children can use it. Even so, it remains important for parents to monitor what children watch, including on this platform.

Primary School Age, Skill Building, Study Support, and Safe Games

Epic: A digital library that includes thousands of illustrated books and children's stories suitable for ages six to twelve. It has received strong ratings and is often described as one of the best reading applications for children. It also includes a read-aloud feature that supports children who are still learning to read. It is a useful way to build a reading habit through enjoyable content.

Quizlet: A learning tool that supports studying and memorization. It is suitable for older children in primary school and early middle school. Children can use it to create flashcards for vocabulary or school terms and practice through small games and short quizzes. It helps strengthen recall engagingly and competitively.

ScratchJr: A free, simplified application that introduces children aged five to seven to basic coding concepts. It allows them to animate cartoon characters and create interactive stories and simple games by arranging visual coding blocks rather than writing text code. It encourages logical thinking and creativity at an early age. For older children, roughly eight to sixteen, they can move to the full Scratch platform on a computer to build more complex projects and share them in a child friendly global community.

Minecraft: A well-known game that is worth mentioning because it can be relatively safe and beneficial when used appropriately. It is a three-dimensional building and exploration game that allows children to express creativity by constructing cities and farms and solving puzzles. It supports single-player and multiplayer use. Parents can set the game to Creative Mode and restrict chat with strangers to support a safer play environment. Some studies have suggested that the game can support learning by developing problem-solving and creativity, and it has been used in some schools to support interactive learning.

Google Arts and Culture: This application can be a valuable resource for children interested in art and history. It allows users to explore famous museums and exhibitions around the world virtually and to learn about artworks and artifacts through interactive features. It is generally more suitable for children aged ten and above, and it can broaden their awareness of global cultures and heritage in an engaging way.

Duolingo: A well-known language learning application that has an approach similar to that of games. Although it is often designed with older learners in mind, children aged ten and above can use it to learn the basics of a foreign language, such as English or French, through levels and challenges that encourage consistency.

There is also a simplified version called Duolingo Kids designed for younger children, with a more child-friendly interface and easier exercises.

Early Adolescence: Creative and Social Applications and Games for Teenagers

Teen-Focused Social Platforms:

Rather than introducing young teenagers immediately to large public platforms dominated by adult users, families may consider alternative social applications that are specifically designed for older children and early adolescents. These platforms typically include stronger safety features and more controlled interaction. One example is Zigazoo, a social application that resembles TikTok in format but is designed for children. It focuses on positive, creative videos and safer engagement. Certain features require parental approval, and the platform offers age-appropriate challenges that encourage creativity and self-expression.

Another example is Groom Social, which supports social interaction through messaging and video sharing within stricter moderation systems. It may include filters that automatically remove offensive language and can require parental verification by email before an account is created. In some cases, parents are also able to monitor activity through a companion application. Platforms such as these provide a more secure social environment and allow teenagers to gradually develop responsible online habits before transitioning to broader platforms as they mature.

Design and Art Tools for Creative Teenagers:

Teenagers with artistic or creative interests can benefit greatly from digital tools that support skill development. More advanced drawing applications, such as Procreate or SketchBook, allow young users to experiment with digital illustration and refine artistic techniques. In addition, simple graphic design and video editing tools, including Canva and iMovie, enable teenagers to produce their own visual or multimedia content. These tools support creativity while also encouraging planning, patience, and problem-solving skills.

Youth-Focused Courses and Workshops:

Teenagers can also be guided toward structured learning opportunities available through digital platforms. With appropriate supervision, platforms such as Coursera and Udemy offer free or low-cost courses in areas such as coding, science, digital design, and game development. Beyond global platforms, local initiatives in the Gulf region increasingly support young digital talent. Youth hackathons, innovation challenges, and technology competitions have become more common. For example, Dubai has hosted student competitions focused on smart application development, where teenagers received recognition for innovative ideas. These initiatives reflect growing institutional support for digitally skilled youth and provide teenagers with meaningful goals to pursue through technology.

General Tips for Choosing Applications and Games

When selecting digital content for children and teenagers, parents should follow several practical guidelines to ensure safety and value.

Use reliable and independent sources for recommendations. Organizations such as Common-Sense Media provide age-based reviews of applications, games, and digital content. Similar guidance is offered by online safety organizations, including National Online Safety.

Read user reviews carefully, especially those written by parents, as they often reveal hidden issues such as unmoderated chat features, inappropriate advertising, or unexpected content.

Whenever possible, test the application personally before allowing a child to use it, or participate with the child during the initial sessions.

Prioritize applications that are free of advertisements and do not request personal data from children. While many educational applications require payment, the cost can often support a safer experience with fewer distractions and reduced commercial pressure.

Make use of in-app safety settings when available. Many games and platforms allow parents to disable chat features, limit friend lists, or restrict interaction with unknown users.

By selecting appropriate applications and games for each developmental stage, families can transform screen time into productive and meaningful experiences. Children and teenagers can enjoy technology while developing knowledge, creativity, and practical skills. However, supervision and guidance remain essential to ensure that digital use supports wellness rather than becoming passive entertainment with little long-term value.

The Effects of Unguided Technology Use: Behavioral, Psychological, and Social Dimensions

Allowing children to engage with technology without guidance or clear boundaries can lead to several negative consequences affecting behavior, mental health, and social development. Numerous studies have examined these risks in the digital age. The following section highlights the most concerning outcomes associated with excessive or unguided technology use among children.

Behavioral Problems and Digital Addiction

Extended periods of unsupervised internet or video game use can, in some cases, develop into digital addiction. Common warning signs include excessive attachment to screens, resistance or distress when devices are removed, and emotional outbursts when limits are enforced. Over time, this behavior may cause children to neglect school responsibilities, family interaction, and physical activity. Research suggests that prolonged screen exposure can reduce opportunities for meaningful engagement within the family environment.

Some reports also point to a connection between heavy social media use and reduced wellness among children. While findings vary, several studies indicate that excessive daily screen use may be associated with lower life satisfaction for some individuals. This raises a broader concern: digital overuse can gradually diminish a child's enjoyment of offline activities, including simple pleasures such as family time or outdoor play. Because children's brains are still developing, constant exposure to fast-paced digital stimulation may weaken attention control and make it more difficult to engage in tasks that require sustained effort, such as studying.

Psychological and Emotional Effects

Unhealthy technology use has increasingly been linked to psychological challenges. Children who encounter violent or frightening content without supervision may experience nightmares, anxiety, or lingering fears. Exposure to online material that promotes harmful values can also create confusion about identity and behavior, particularly when children lack adult guidance.

Some research suggests that even the mere presence of a smartphone during a concentration task—regardless of whether it is actively used—can reduce attention and cognitive performance. This constant awareness of a nearby device may keep the mind in a state of anticipation, interrupting deep focus and affecting academic performance. In addition, excessive nighttime device use can interfere with sleep patterns. Many children who use screens late into the night report difficulty falling asleep or staying asleep, which can lead to irritability, mood swings, and emotional instability during the day.

Social Isolation and Weakened Social Skills

Although technology is designed to connect people, excessive and unguided use can sometimes produce the opposite effect. A child who spends most of the day engaged with a device may lose valuable opportunities for face-to-face interaction with family members and peers. Over time, this can weaken essential social skills such as conversation, emotional expression, and active listening. Without early intervention, some children may begin to prefer isolation and withdrawal from real-world social settings.

Studies frequently report that heavy internet use can reduce family communication and limit participation in physical, cultural, and community activities that support healthy development. For adolescents, an additional risk lies in replacing meaningful friendships with superficial online connections. This can contribute to feelings of emptiness or loneliness, even when a teenager appears socially active through numerous online contacts.

Exposure to Inappropriate Content and Ethical Risks

The internet is an open and largely unfiltered space. When a child explores it without guidance or supervision, they may easily encounter content that is unsuitable for their age and stage of development. One of the most serious dangers lies in the internalization of unhealthy ideas and behaviors. Exposure to programs, videos, or games that promote aggression, distorted social norms, extreme violence, or sexual content can influence a child's understanding of right and wrong and disrupt a healthy sense of innocence. Over time, such exposure may contribute to aggressive tendencies, unjustified fears, or inaccurate beliefs. For example, a child may begin to view the world as inherently frightening and filled with conflict after repeated exposure to violent or chaotic digital narratives. Some games also normalize hostile language and aggressive reactions, which children may imitate in real-life interactions with peers.

From an ethical and values-based perspective, certain online materials promote ideas that may directly conflict with a family's religious, cultural, or moral framework. When children encounter such content without discussion, explanation, or contextual guidance from parents or caregivers, they may absorb these messages uncritically. Without a clear reference point, children can become confused about values and acceptable behavior, particularly at an age when their moral understanding is still forming.

Physical Health Problems

Extended periods of screen use, especially when combined with limited physical movement, are associated with several physical health concerns. These include weight gain, eye strain, and back or neck pain resulting from poor posture. Children who spend long hours engaged in fast-paced digital games may also develop unhealthy eating patterns, such as overeating while using screens or skipping meals altogether, leading to irregular eating habits. Continuous exposure to screens can cause eye fatigue and dryness, while blue light may interfere with the body's natural sleep rhythms. Collectively, these factors contribute to reduced physical activity, persistent tiredness, and low energy levels, which can affect a child's overall health and daily functioning.

Reinforcing Violent or Harmful Behavior

Some video games and widely shared online clips reinforce negative values, such as resolving conflict through violence, reacting with anger, or mocking others for entertainment. When young children are repeatedly exposed to scenes of fighting, shooting, or verbal aggression, they may become desensitized to violence and begin to view it as normal or acceptable. In certain cases, children attempt to imitate these behaviors with classmates at school, which can contribute to increased aggression and disciplinary problems.

Similarly, repeated exposure to frightening, pessimistic, or emotionally charged content can contribute to ongoing anxiety or a negative worldview. Several studies have reported associations between children's social media use and lower life satisfaction, particularly in relation to self-confidence, body image, and perceptions of family relationships. These findings suggest that constant comparison with curated online images and lifestyles can weaken self-esteem and encourage unhealthy behaviors as children attempt to resemble unrealistic ideals presented online.

Exposure to Exploitation and Cybercrime

One of the most serious and alarming risks children face online is sexual exploitation or blackmail by strangers. Offenders can conceal their identities behind screens and often target children through online forums, multiplayer games, or chat applications. Various reports have presented troubling figures, indicating that a significant proportion of children across different countries have felt at risk of sexual exploitation online. While estimates vary across studies and regions, the central reality remains unchanged: this threat exists and requires proactive prevention.

Local reporting has documented cases in which online predators deceived children into sharing private images and later used them for blackmail. Such experiences can be deeply traumatic and damaging to a child's psychological wellness if not detected and addressed quickly. A strong digital wellness framework seeks, above all, to prevent children from reaching these dangerous situations through awareness, supervision, and the development of strong safety habits.

Real incidents further emphasize the importance of vigilance. For instance, in Dubai, a man was sentenced to three years in prison for using Snapchat and WhatsApp to lure children into exchanging inappropriate images. The case came to light when a mother noticed suspicious messages on her teenage son's phone and immediately reported them to the authorities. This example demonstrates how parental attentiveness can interrupt harm at an early stage. Recognizing warning signs—such as sudden withdrawal, secretive behavior, or concealed digital activity—is a critical element of effective child protection in the digital age.

In summary, unguided technology use is comparable to allowing a child to walk into a crowded street without supervision or instructions. The likelihood of harm increases significantly. In contrast, when children receive consistent guidance and appropriate supervision, the internet can become a space that is safer to explore and learn from. Protecting children online is a shared responsibility involving families, schools, and institutions, yet the most direct and continuous role begins at home through guidance, care, and open communication. Encouragingly, many negative effects can be prevented or reduced by applying the digital parenting strategies discussed throughout this chapter. When challenges do arise, early detection and timely intervention can protect children from more serious harm. For this reason, families should remain attentive to children's digital behavior—not out of suspicion, but out of care and commitment to their psychological and social wellness in a rapidly evolving digital world.

Real Stories from the Gulf and the World: Successes and High-Risk Situations

Real-life examples help transform abstract concerns about digital wellness into concrete and relatable experiences. In this section, we present stories from the Gulf region and beyond that highlight both inspiring successes and cautionary situations. These cases offer practical lessons and demonstrate how digital challenges can be addressed constructively.

A Success Story from the Gulf: The Young Innovator

In 2021, a Qatari student received national recognition for an initiative aimed at addressing bullying, including cyberbullying, within the school environment. The story began when the student noticed a rise in online bullying among classmates and decided, while still in primary school, to take action. Working in collaboration with the school, the student helped organize awareness workshops focused on the risks of cyberbullying and safe ways to respond. In addition, the student proposed a simple digital reporting system that would allow students to report bullying incidents confidentially and connect directly with a school social counselor.

This example illustrates how children growing up in a technology-driven generation can use digital tools for positive social impact when they receive encouragement and guidance. It also highlights the importance of involving young people in designing solutions to the digital challenges they experience firsthand, while adults provide structure, support, and oversight.

A Global Success Story: A Teen App Developer

Globally, there are also inspiring examples of teenagers using digital skills to support education and community needs. In one case, a young teenager in India developed a simple mobile application that helps children learn English vocabulary through interactive, game-based activities. The teenager developed an early interest in coding and built skills through online learning resources. During the period when schools were disrupted by the COVID-19 pandemic, the teenager used additional free time to create an educational tool to support younger siblings and peers. The application later gained local recognition within youth innovation circles and became a motivating example for other students interested in learning programming and creating useful digital products.

This story reinforces a crucial point: digital wellness is not solely about avoiding risks. It also involves nurturing children's technical creativity and empowering them to become capable, responsible, and constructive digital creators who contribute positively to their communities.

A Cautionary Story from the Gulf: The Blue Whale Challenge

On the other hand, there are painful stories that reveal the serious risks children may face in digital spaces. In 2018, media reports in Saudi Arabia described two tragic child deaths that were linked to an online phenomenon known as the Blue Whale Challenge. This challenge reportedly targeted young users, particularly adolescents, through a series of dangerous and gradually escalating tasks spread over approximately fifty days. The tasks were described as beginning with disturbing content or self-harm–related prompts and ending with instructions that could lead to fatal outcomes.

The Blue Whale Challenge was widely reported internationally, raising concern in several countries. In Saudi Arabia, public responses included awareness messages issued by media outlets and health-related institutions, warning families about the existence of such hidden online challenges. This case demonstrates how covert digital trends can draw children in through curiosity, emotional vulnerability, or social pressure. When there is no early awareness or timely intervention, the consequences can be devastating. It also serves as a powerful reminder that supervision and honest, open conversation with children can play a protective role—sometimes in the most literal sense.

A Cautionary Global Story: Online Blackmail

In a widely reported case in Europe, a teenage girl communicated online with someone she believed to be a peer. Over time, the interaction developed into the sharing of private images, which she assumed would remain within a trusted relationship. Instead, the other person became an online blackmailer, threatening to publish the images unless she paid money or provided additional content.

The girl suffered in silence due to fear, confusion, and shame, and her mental wellness declined rapidly. Fortunately, her family noticed changes in her behavior and signs of distress and insisted on understanding what was happening. She eventually shared the truth with them, after which they contacted the authorities. Through digital investigation, the offender was identified and arrested, and it later emerged that he was an adult rather than a teenager.

Although distressing, this story carries essential lessons. Children and teenagers need clear education about the risks of sharing private images or sensitive personal information, even with people they believe they know. They also need reassurance that they can turn to their parents without fear or punishment, even when a mistake has been made. This case connects closely with recent awareness efforts in the Gulf addressing online blackmail, including the Gulf Health Council Campaign What Should Not Be Silenced, which focuses on child exploitation and outlines practical prevention steps.

A Positive Story: A Youth Initiative for Digital Awareness

In contrast, there are also encouraging stories that show how young people themselves can contribute to safer digital environments. A group of high school students in the United Arab Emirates formed a school club and launched a digital awareness campaign titled "Let Us Be Safe Online." They created short videos and shared them on platforms such as Instagram and TikTok. The videos presented acted scenes showing realistic online risks faced by teenagers, including messages from strangers requesting personal data or friends encouraging participation in dangerous online challenges. Each video concluded with a practical safety tip or a correct response strategy.

The videos spread widely among students in other schools and received positive feedback from parents and teachers. The students were later recognized by the Ministry of Education for an initiative that highlighted the active role young people can play in promoting digital awareness. This story demonstrates how the lived experiences of the digital generation can be transformed into effective peer guidance, delivered in language and formats that resonate strongly with young audiences. It also reflects the value of digital parenting approaches that nurture responsibility, leadership, and social awareness.

Together, these real stories teach a central lesson: technology is a double-edged tool in the lives of children. It can support creativity, learning, and achievement, but it can also expose them to serious harm when guidance is absent. Success stories offer hope and motivation to continue investing positively in children's digital development, while cautionary stories remind families to remain alert and informed.

For parents and educators, sharing these stories with children in age-appropriate ways can be particularly powerful. When a child learns that someone else faced serious consequences because of a reckless online challenge or misplaced trust in a stranger, they may pause before making a similar choice. When they hear about peers using technology to achieve positive outcomes, they may feel inspired to follow that path. Stories carry strong educational value and can be an effective tool for strengthening children's digital wellness.

How We Teach Children to Protect Themselves Digitally with Awareness and Independence

The highest goal of digital parenting is to help children develop a level of digital awareness that enables them to protect themselves independently when they are online. Parents cannot monitor every click, message, or interaction forever. For this reason, children need an internal form of protection—knowledge, judgment, and confidence—that supports safe decision-making in digital spaces. The following strategies aim to help children learn how to protect themselves online with awareness and growing independence.

Ongoing Education About Online Risks and Prevention

Parents should provide children, using simple and age-appropriate language, with clear information about the risks they may encounter online and how to respond safely. For example, explain what malware and computer viruses are, and clarify that some links can harm a device or steal personal data. Teach children not to click on unknown or suspicious links. Explain what cyberbullying is and make it clear that anyone who harasses or sends abusive messages is in the wrong. Emphasize that the correct response is to inform a trusted adult and block the person immediately. Introduce the concept of impersonation, explaining that someone online may pretend to be a child while actually being an adult. For this reason, children should not place trust in people they do not know in real life.

Knowledge itself is a form of protection. The more a child understands risky situations in theory, the more capable they become of recognizing them in practice and avoiding harm. Observations from educational settings suggest that children who receive direct awareness training often respond more effectively in risky situations. For instance, in one school in Sharjah, police officers delivered sessions to middle school students on recognizing cyber threats, understanding the importance of informing parents, and checking links before opening them. After these sessions, one student reported that her mother began discussing online risks regularly and that they started verifying suspicious content together. This shared awareness and cooperation reflect exactly what effective digital parenting seeks to achieve.

Practicing Through Role Play and Hypothetical Scenarios

One practical and effective method is to role play with your child by asking "what if" questions. For example, you might ask, What would you do if you received a message from someone you do not know saying, "You have won a prize. Click this link to claim it"? Allow the child time to think and respond, then gently guide them toward the safer choice and explain the reasons behind it. You can also ask, What if a classmate sends you a funny video link? Do you click it immediately? Use this moment to discuss the possibility that the account could be compromised or that the link itself could be suspicious. These simple training scenarios help children practice decision making for moments when a parent is not nearby.

Parents can also use child focused awareness videos that present similar situations. Some educational animated programs, available in both Arabic and English, show cartoon characters encountering online risks and learning how to respond appropriately. After watching together, ask your child reflective questions such as, What mistake did the character make at first? How did they correct it later?

What would you do in the same situation? These conversations help transform general advice into practical, memorable understanding.

Reinforcing and Respecting Digital Privacy

Teach your child that personal data has value and should be protected online. Begin by asking what personal data means, then explain that it includes a full name, home address, phone number, private photos, school name, and similar identifying details. Emphasize that no one should request this information online, and that if anyone does, the child should inform you immediately.

You can also complete a simple activity together. Create two lists: one for information that can be shared publicly, such as hobbies or a favorite game, and another for information that should never be shared, such as family details, addresses, or private photos. In addition, parents should act as role models. Avoid sharing your child's sensitive information or images on social media without a genuine need. Many children express discomfort when parents post their photos without permission, so showing respect for their digital privacy sends a powerful message. At the same time, teach children to respect the privacy of others. They should not post a friend's photo without consent or share another person's private information. These practices are a core part of ethical digital safety.

Building a Child's Digital Conscience

Alongside external rules, children need an internal moral compass that guides their online behavior. Reinforce the values your family believes in and explain how these values apply in digital spaces as well. If your family values honesty, discuss how this translates online, such as not lying about age or identity and not cheating in games. If your family values respect, explain that this includes avoiding harmful comments, insults, or rumors about others. As one specialist observed, the behavior and values we expect from children in real life should also apply on the internet.

For example, if you expect your child to be kind and polite offline, the same expectations apply on social media and gaming platforms. Children should not participate in cyberbullying or mock others. When children understand that the internet is not a separate world without rules, but an extension of everyday life governed by the same ethical standards, they are more likely to behave responsibly even without direct supervision.

Building Resilience and Emotional Intelligence When Facing Digital Challenges

One of the most important skills parents can teach is digital resilience—the ability to face online problems calmly without feeling overwhelmed or defeated. Explain to your child that mistakes can happen. A child may download an unsafe application accidentally or reply to a message and later regret it. These situations are possible and normal. What matters is not reacting with fear or secrecy, but learning from the experience and moving forward. Reassure your child that, as a family, you will face any digital problem together.

A widely accepted parenting principle is that the best way to build online resilience is to create an environment where children feel safe talking about anything they encounter. If your child makes a mistake online, respond calmly, listen carefully, and focus on solving the issue together rather than delivering a long lecture. This approach increases the likelihood that the child will speak up quickly in the future. Parents can also teach simple stress management strategies. If a child encounters upsetting content, they can close the screen, take slow deep breaths, and then come to you to talk about what they saw.

Using Protective Tools Together with the Child

Introduce children to basic digital safety tools and practice using them together. Show them how to block someone who sends abusive messages or unfamiliar contact requests. Teach them how to report inappropriate content on platforms such as YouTube or TikTok. Explain that search engines offer Safe Search settings and that enabling them helps filter harmful results. Practicing these steps builds confidence and helps children feel actively involved in protecting their own safety.

Some regional reports suggest that a significant number of children have been contacted by strangers online. However, children who know how to block users and strengthen privacy settings can stop unwanted contact quickly. Teaching these skills early ensures children know what to do in uncomfortable situations. You can also explain antivirus software in simple terms, describing it as a protective shield for devices, which helps children understand that cybersecurity is something they can actively participate in.

Gradually Increasing Independence and Showing Trust

When a child demonstrates consistent understanding and respect for safety rules, parents can gradually increase digital independence. For example, if a child has used devices responsibly for several months, you might allow limited internet browsing alone in a shared space such as the living room. If a child is around twelve and you decide to allow a first social media account, such as a private Instagram profile, create it together, adjust privacy settings together, and then allow the child to manage it with light supervision. Explain clearly that this step reflects trust in their judgment.

This approach strengthens a child's sense of responsibility. Supervision remains important, but it should not be so restrictive that it removes opportunities for growth. The central idea is that children should learn safe digital independence gradually, much like learning to ride a bicycle and later to drive a car. At first, adults stay close. Then they step back while offering guidance from a distance. Over time, children develop the confidence and skills to navigate digital spaces responsibly on their own.

A child's ability to protect themselves online is the result of education and empowerment. Families should teach core principles, build awareness step by step, and give children chances to apply what they have learned in safe situations. When a child or teenager reaches a point where they know what to do and what to avoid, and can make the right choices even without parents present, we can confidently say they are on a strong path toward lasting digital wellness.

Gulf and Global Initiatives to Protect Children Online

Governments and institutions around the world, and particularly across the Gulf region, have increasingly recognized the need to act decisively to protect children in cyberspace. As children's digital access expands, so too does the responsibility to ensure that online spaces remain safe and supportive. In response, many initiatives and programs have been launched to raise awareness among children and parents, and to establish policies and safeguards that strengthen digital safety. The following sections highlight several prominent efforts from the Gulf and the international community.

Gulf Initiatives

"The Child Digital Safety Initiative in the United Arab Emirates"

In March 2018, institutions in the United Arab Emirates announced a comprehensive national initiative titled "Child Digital Safety," known in Arabic as "Digital Safety for Children." The initiative aims to raise awareness among children and school students about online risks, while encouraging safe and positive internet use. One of its key strengths is that it does not focus on children alone. It also provides parents and teachers with practical guidance and tools that support children's digital safety in everyday settings.

The initiative included a digital awareness portal offering educational materials and support tools, as well as interactive camps for children aged five to eighteen. In these camps, online safety concepts were introduced through age-appropriate and engaging activities. The program also offered training workshops for parents and teachers, alongside a dedicated support channel to respond to urgent family questions about digital safety. Over time, this effort was reinforced by additional projects aimed at strengthening cybersecurity practices in schools, sometimes framed under themes such as creating a safe school environment.

National reports have suggested that a proportion of children encounter digital threats, and that awareness programs play an important role in reducing risk by improving early recognition and encouraging safer behavior. The broader lesson is clear: structured education, combined with family involvement, significantly strengthens children's ability to navigate online spaces with confidence and caution.

"Awareness Campaigns in Saudi Arabia and Other Gulf Countries"

In 2022, the Gulf Health Council launched a regional awareness campaign titled What Should Not Be Silenced. The campaign addressed sensitive issues, including online blackmail involving children, and aimed to educate both parents and young people about what digital blackmail is, how to recognize warning signs, and how to prevent or respond to it. The messaging was simplified, evidence informed, and developed with specialist input, then delivered through media outlets and social platforms to reach families effectively.

In Saudi Arabia, official bodies have also introduced awareness resources for families and educational content on cybersecurity and online safety through national learning platforms. The National Cybersecurity Authority and related institutions have supported training and awareness initiatives, including programs for teachers, to strengthen digital literacy and safe online conduct among students. In Bahrain, reporting mechanisms have been established to receive and follow up on cases related to online child exploitation, supported by public awareness efforts delivered in partnership with community organizations.

In Oman, Kuwait, and Qatar, schools and local media have similarly hosted programs that promote positive internet use and explain how young people can protect personal data and avoid unsafe interactions with strangers. Across these countries, a shared goal emerges: to frame children's online safety as a collective responsibility, supported by education, clear reporting pathways, and practical tools for families and schools.

"The School-Based Digital Safety Ambassadors Initiative"

Many schools across the Gulf region have adopted an innovative peer-led approach by training students to serve as digital safety ambassadors. For example, a school in Dubai introduced a program that prepares older students to guide and support younger peers on online safety issues. These ambassadors communicate messages in age-appropriate language and formats that resonate with students.

This approach not only strengthens awareness but also builds a culture of responsibility and peer support within the school community. Education authorities have encouraged this model as a promising practice that could be expanded more widely if proven effective.

Global Initiatives

"United Nations Guidance"

In 2020, the International Telecommunication Union (ITU), a United Nations agency, released global Child Online Protection Guidelines. Developed with input from experts across multiple sectors, the guidelines provide resources tailored to different audiences, including children, parents and educators, technology companies, and policymakers. They outline international best practices, practical advice, and adaptable policy models that countries can localize according to their needs. The ITU has also supported implementation through online training resources aimed at strengthening children's online safety worldwide.

"Safer Internet Day"

Safer Internet Day is an annual global event observed each February in more than one hundred and seventy countries, including several Arab and Gulf states. Schools, community organizations, and youth groups use the day to promote responsible internet use and raise awareness about digital risks. Although the theme changes each year, the central message remains the same: creating a safer internet is a shared responsibility. In the Gulf, many schools mark the occasion through student-led activities, posters, presentations, and creative performances that make online safety relatable and engaging.

"The We PROTECT Global Alliance"

The WePROTECT Global Alliance is a global coalition that brings together governments, major technology companies, and civil society organizations to prevent and combat online child sexual exploitation. The alliance supports cross-border information sharing and promotes research and technological tools that help identify and stop offenders online. Several Gulf countries, including the United Arab Emirates and Saudi Arabia, participate in this alliance and contribute to the exchange of legal and technical best practices.

"Technology Company Initiatives"

In the private sector, major technology companies such as Google, Meta, and Microsoft have introduced initiatives aimed at protecting children and teenagers online. Google, for example, developed the educational program Be Internet Awesome, which uses interactive lessons and games to teach children core online safety principles in an engaging way. In Arabic-speaking contexts, Google has also provided localized resources, including Arabic safety materials and an educational game that introduces concepts such as privacy, security, and respectful online behavior.

Meta has partnered with regional institutions on awareness efforts related to cyberbullying and online exploitation, using resources distributed through its platforms. It has also strengthened protections for teenagers on Instagram, including default privacy settings for younger users and limits on messages from unknown adults, depending on age verification and account settings.

YouTube offers a child-focused environment through YouTube Kids and also maintains systems for reviewing content intended for families. These systems are designed to reduce children's exposure to inappropriate material while allowing access to age-appropriate videos. Similarly, TikTok has introduced a Family Pairing feature that allows a parent to link their account to a teenager's account and manage selected settings, such as screen-time limits, direct messaging controls, and content visibility. Features like these reflect a growing acknowledgment by platforms of their role in supporting family-led digital supervision.

without guidance and boundaries, children's digital environments can become chaotic and unsafe.

"We teach all our students the importance of cybersecurity, the risks of online interactions, and the positive opportunities the digital world can offer when used responsibly."

~ Kari Houza, Deputy Principal and Child Protection Lead, Dubai

This quotation reflects the voice of educators working directly with students. It illustrates how leading schools adopt a comprehensive approach that balances awareness of risks with encouragement of responsible and creative digital use.

"Let your child know that you will love them no matter what, and that any problem, whether online or offline, is something you can overcome together."

~ Advice from the Tech Ye website, attributed to a family technology specialist

This statement provides a powerful conclusion. It emphasizes that unconditional support and reassurance form the foundation of a healthy parent–child relationship. When children feel emotionally safe, they are far more likely to seek help when mistakes occur or risks arise. This sense of trust is one of the strongest protective factors in digital parenting.

Together, these expert insights serve as guiding principles for families and educators. They offer a moral and practical compass for everyday decisions in raising children in the digital age. By learning

Legislation and Laws to Protect Children's Digital Privacy

Many countries have also introduced laws specifically aimed at protecting children's digital privacy. In the United States, the Children's Online Privacy Protection Act (COPPA) restricts the collection of personal data from children under the age of thirteen without verified parental consent. In the European Union, the General Data Protection Regulation (GDPR) includes special provisions governing the processing of minors' personal data, placing stronger obligations on organizations to safeguard children's information.

In the Gulf region, the United Arab Emirates has included explicit digital protections within its child rights legislation, commonly known as Wadeema's Law. These provisions require telecommunications companies and internet service providers to report the circulation of child sexual abuse material, thereby strengthening legal accountability and deterrence. In addition, many countries have established reporting hotlines for online child abuse and exploitation, making it easier for families and professionals to alert authorities and trigger a rapid response.

Taken together, these initiatives and legal frameworks send a clear message: protecting children online is a shared responsibility. Governments contribute through legislation and public awareness campaigns, schools contribute through education and values building, technology companies contribute by developing safer platforms and tools, and families contribute through supervision and guidance. Despite ongoing challenges, it is encouraging to see Gulf countries playing a leading role in regional efforts. For example, the United Arab Emirates has developed national cybersecurity strategies that include explicit attention to child protection, while Saudi Arabia has positioned cybersecurity as part of its long-term national vision, with programs that specifically target younger generations. This growing focus, both locally and globally, means that more resources are becoming available for parents and educators, and it suggests that sustained collaboration can support a safer digital environment for children over time.

The Role of Schools and Teachers in Supporting Children's Digital Wellness

Building digital wellness is not limited to the home environment alone. Schools and educational settings play a crucial and complementary role in shaping children's safe, informed, and responsible use of technology. In practice, teachers and school leadership often act as a second line of protection after the family, helping to guide digital behavior and respond when risks arise. The following points outline key aspects of the school's and teacher's role in supporting children's digital wellness.

Integrating Digital Citizenship into the Curriculum

Schools should teach the principles of digital citizenship either as part of the formal curriculum or through structured classroom activities. This includes lessons on digital privacy, online fraud, cyberbullying, the rights and responsibilities of digital users, and respect for intellectual property. In some countries, these topics have been integrated into technology courses or student guidance programs. In the United Arab Emirates, for example, there has been a growing emphasis on embedding digital education within school curricula so that students learn, from an early age, the ethical standards and boundaries required for responsible engagement with modern technology. When children receive this knowledge at school, it reinforces what they learn at home and strengthens their overall understanding.

Training and Empowering Teachers

Teachers themselves need appropriate training so they remain informed about evolving digital risks and effective response strategies. In the past, many educators did not receive systematic preparation in this area, but this situation has been gradually improving. Education authorities across the region have introduced workshops and training programs for teachers and school counselors focused on cybersecurity, online safety, and student wellness. In Qatar, for example, initiatives have emphasized anti-bullying and digital awareness training for teachers, enabling them to provide informed guidance. In Bahrain and Saudi Arabia, student support and counseling staff have also been trained to recognize signs of online abuse and to respond appropriately.

A teacher who understands contemporary digital culture is better positioned to notice early warning signs affecting students, whether these involve academic decline linked to excessive gaming or social withdrawal connected to online harassment.

Establishing Clear School Policies for Device Use

Many schools now provide tablets or computers as part of the learning process, or allow limited student use of personal devices. This makes clear and consistent technology-use policies essential. For example, schools may prohibit phone use during lessons unless explicitly permitted by the teacher for educational purposes. They may also restrict access to certain websites through the school network by using security filters and protective systems.

Schools can further promote responsible behavior by teaching classroom device etiquette, such as not recording others without consent and not posting school-related content without permission. These policies protect both students and staff while helping young people develop digital self-discipline. Some leading schools go a step further by introducing a school digital charter that students and parents sign at the beginning of the academic year. Such a document clarifies digital rights and responsibilities within the school's learning environment and sets shared expectations.

Monitoring and Identifying Students' Digital Behavior at School

Teachers and computer lab supervisors should monitor students' online activity when devices are used on school premises. For instance, if a teacher observes a student accessing inappropriate websites or engaging with violent games, the issue should be addressed through guidance, and parents may need to be informed if the behavior persists. Schools should also remain alert to signs of cyberbullying, such as reports of harassment within class-based messaging groups, and respond promptly with clear educational and disciplinary measures.

Many Gulf schools now offer confidential reporting channels, such as dedicated hotlines or email addresses managed by student guidance services. These channels encourage students to speak up about concerns without fear, reinforcing a school culture where digital safety is taken seriously and addressed proactively.

Running Regular Awareness Workshops and Events

Schools can play an active role by organizing regular digital awareness workshops and events. These may include digital safety days during which specialists or trained police officers speak with students using clear, age-appropriate language. In one media interview, a student described how a school in Sharjah hosted a session led by police officers on recognizing online risks and understanding the importance of informing parents. According to the student, classmates benefited greatly from the session and became aware of issues they had not previously understood.

Schools can also encourage student participation by organizing projects focused on safe internet use. Examples include exhibitions displaying student-designed posters and leaflets, or competitions for the best short digital awareness video created by students. Activities such as these transform digital safety from an abstract concept into a practical and engaging learning experience, helping students internalize lessons more effectively than through theory alone.

Involving Parents and Coordinating Efforts

Schools cannot address children's digital wellness in isolation from the home environment. Teachers and school counselors should maintain consistent communication with parents regarding students' digital behavior and wellness. For instance, if a student appears persistently tired or sleepy, possibly due to late-night device use, a school counselor can contact the family to discuss the issue constructively. Similarly, if an incident of cyberbullying occurs between students outside school hours, the school should inform the relevant families and coordinate a response that prioritizes safety, accountability, and learning for all involved.

Some schools share regular digital safety tips with parents through newsletters, including warnings about risky online challenges or harmful games that may be circulating. Parent councils and school meetings may also host specialists who explain practical monitoring tools and effective home strategies. This level of coordination creates a consistent message for the child across settings. When school and home reinforce the same guidance and values, children are far more likely to internalize safe and responsible digital behavior.

Supporting Positive Uses of Technology in Education

Schools should avoid focusing solely on digital risks. Equally important is guiding students toward the positive and constructive opportunities technology can offer. A skilled teacher can channel students' interest in technology into meaningful educational activities, such as online research projects, collaborative assignments, or classroom tasks using interactive learning applications. When students feel that teachers trust them to use technology productively, they are more inclined to protect that trust and behave responsibly.

One school leader in Dubai described this balanced approach by noting that schools recognize the growing challenges children face in the digital world and remain committed to equipping them with the knowledge and skills needed to stay safe. At the same time, students are taught about the positive opportunities technology can provide when used responsibly. This balance is essential.

Technology should not be presented as a threat in itself, but rather as a powerful tool that requires wisdom and self-control.

In summary, schools are a vital partner in building children's digital wellness. They are environments where children spend many hours each day and are exposed to a wide range of social influences. When schools educate students effectively, monitor digital behavior, and provide consistent guidance, the impact extends beyond the classroom and contributes to broader community awareness. It is also important to remember that teachers serve as influential role models. A teacher's words about safe internet use may sometimes leave a deeper impression than those of parents. For this reason, equipping teachers with knowledge and practical tools is a necessary investment. When home and school work together, they create a protective and supportive system around the child that reduces digital risks and helps prevent many online difficulties.

Quotes from Specialists in Education and Technology

When discussing children's digital wellness, it is useful to recall insights shared by experts in this field, as these perspectives reflect years of research and professional experience. Below is a selection of quotations frequently cited by specialists in education and technology regarding raising children in the digital age.

"Raising children in the digital age is not easy."

~ UNICEF

"Social media and smartphones are advancing at a pace that seems faster than our children's development, and it is not easy for parents to keep up. However, through guidance and awareness, we can ensure that our children enjoy the benefits of technology in a safe and responsible way."

This quotation, drawn from UNICEF's work on digital parenting, reflects a reality faced by many families and educators. Technology evolves rapidly, often faster than children mature. At the same time, the message remains hopeful, informed guidance can help transform digital use into a positive and safe experience.

"Like most things in life, technology can be used for good or for harm. As a parent, you can help your child understand the difference."

This statement captures the balanced philosophy of effective digital parenting. Technology should neither be demonized nor left without limits. Instead, parents play a key role in helping children make thoughtful choices and use technology for positive growth.

"Children learn through observation and imitation. Parents should be good role models in how they use technology."

This reminder reinforces a timeless educational principle, children learn by watching adults. Responsible digital behavior must therefore be demonstrated consistently, not merely instructed.

"When law is absent from any society, it declines. When digital parenting is absent from the digital community, disorder and harmful behavior spread."

This comparison highlights the role of digital parenting as a framework that organizes behavior in online spaces. It warns that

from specialists and applying recommendations grounded in experience and research, we move closer to the goal of raising a digitally aware, balanced, and safe generation.

Analyzing Common Parenting Mistakes in Managing Children and Technology

Despite good intentions, parents and educators may sometimes make mistakes that lead to outcomes opposite to what they desire. This section examines some of the most common errors in managing children's technology use, explains why they can be harmful, and suggests healthier alternatives.

Excessive Restriction and Total Bans

Some families, driven by fear, adopt a strict policy of total prohibition toward digital devices and the internet. They may forbid nearly all technology use beyond school requirements and treat digital tools as entirely off-limits. While the motivation—protection—is understandable, this approach often proves ineffective over time. A child who is completely deprived may become more curious and determined to access technology secretly, without guidance or supervision. This hidden use can increase risk rather than reduce it.

Total bans may also create feelings of unfairness, as children observe peers enjoying what they are denied. Over time, this can contribute to resentment, rebellion, or emotional distance between children and parents. As many specialists note, viewing technology as entirely harmful and banning it completely is rarely a successful long-term strategy.

A healthier correction lies in balance rather than absolute restriction. This involves allowing guided, age-appropriate use, selecting suitable content, and setting clear boundaries instead of enforcing blanket prohibitions.

Excessive Leniency and Neglect

At the opposite extreme, some parents provide children with digital access without meaningful limits or consistent follow-up. This may occur because parents assume their child will not engage in risky behavior, or because they wish to avoid the conflict that often accompanies setting boundaries. A parent may say, "My child is too smart to get into trouble," or, "My daughter is well behaved and I know her," and then provide connected devices without sufficient guidance or monitoring. In reality, any child, even one who appears mature and responsible, can be targeted by skilled online manipulators or exposed to harmful content. Neglecting a child's digital life creates gaps that others may exploit, or that the child may enter through curiosity.

In some families, parents later discover that a child experienced online blackmail for months, or severe cyberbullying, without their knowledge, simply because there was no ongoing dialogue or supervision. This illustrates that excessive confidence combined with a lack of oversight is a serious mistake. Awareness and maturity develop gradually, and children continue to need support and guidance, even when they seem capable of managing technology on their own.

Using Devices as an Unbalanced Reward or Punishment

Many families rely heavily on-screen time as a bargaining tool in daily parenting. A child may receive a tablet or extra screen time as a reward for good behavior, while device access is removed as punishment after mistakes. Although setting time limits is important, the problem arises when devices become the primary tool of pressure and control. This approach can make technology appear even more desirable in the child's mind, transforming it into the "ultimate reward" and increasing fixation on it. Similarly, removing devices for every minor issue may lead to defiance, frustration, or resentment.

A more effective approach is to adopt a balanced system of rewards and consequences that does not depend solely on technology. When technology-related consequences are applied, they should be proportional and directly connected to the behavior. For example, if a child misuses a device by visiting a prohibited website, it is reasonable to restrict access for a defined period. In contrast, removing a device simply because academic performance has declined may be ineffective, as the underlying cause could be learning difficulties rather than excessive screen use.

Lack of Agreement Between Parents and Conflicting Messages

Another common mistake occurs when parents send inconsistent or conflicting messages about technology rules. One parent may enforce a limit of two hours per day, while the other allows four hours, or purchases a device despite the partner's objections. One parent may insist on delaying a personal phone until late adolescence, while the other gives in earlier due to pressure. Such inconsistency weakens parental authority and encourages children to search for loopholes.

To avoid this, parents are advised to agree in advance on a shared approach and present a unified policy to the child. Differences of opinion should be discussed privately, not in front of children. If a technology-related consequence is applied, the other parent should not reverse it without discussion. Consistency is essential for household rules to be respected and effective.

Neglecting Dialogue and Failing to Explain

In some cases, parents impose strict rules without explaining the reasoning behind them. Statements such as, "You cannot play this game, end of discussion," may lead to temporary compliance, but they rarely produce understanding. When adults are absent, children may experiment out of curiosity, especially if they were never given a clear explanation. Failing to explain risks in an age-appropriate manner reduces the effectiveness of guidance.

By contrast, calm and thoughtful explanations increase cooperation. For example, a parent might say, "I am not allowing this game because it is very violent and may affect your mood and stress levels. There are other games that are enjoyable and more beneficial, and we can try them together." Or, "Using your phone before sleep keeps your brain alert and makes it harder to rest, which is why we have a rule of no devices after nine in the evening." When children understand the logic behind rules, they are more likely to cooperate. Even if they do not fully agree, they are more likely to recognize that the rules are intended to protect them rather than control them unfairly. Open dialogue also builds trust and mutual respect.

Being Distracted by Technology and Modeling Poor Habits

Specialists frequently emphasize that parents' technology habits serve as a powerful model for children. A common mistake occurs when parents remain absorbed in phones or social media for extended periods in front of their children. When a child seeks attention and is repeatedly ignored or postponed, a harmful message is conveyed, that the device is more important than the child. This behavior also normalizes excessive screen use and increases the likelihood of imitation.

Additionally, some parents unintentionally violate children's digital privacy by sharing their photos or personal stories online in pursuit of social engagement. This can cause embarrassment or resentment, particularly during adolescence. If parents wish to reduce digital overuse in children, they must avoid modeling the same behavior. If they expect children to respect digital privacy, they must respect it themselves. When parents demonstrate balanced technology use and prioritize family interaction, children are more likely to develop similar habits.

Responding Only After a Problem Instead of Preventing It

Another frequent mistake is adopting a reactive rather than preventive approach. Some families only address digital safety after a serious issue has occurred. A parent may accidentally discover concerning messages on a child's phone and react with panic, or notice behavioral changes linked to cyberbullying only after harm has intensified. Prevention, however, is generally more effective than delayed intervention.

Teaching basic digital safety principles early, observing small warning signs such as mood changes or withdrawal after device use, and responding promptly can prevent more serious problems from developing. In many cases, children who became involved in high-risk online situations had not received guidance beforehand, and they did not feel confident that their parents would respond with understanding if they spoke up. The mistake lies in waiting. The better approach is early and continuous engagement from the beginning.

By avoiding these common parenting mistakes, families can remove many obstacles that weaken digital guidance. Mistakes are a natural part of parenting, and no approach is perfect. However, awareness of these pitfalls allows families to adjust more quickly and effectively. Digital parenting is still a relatively new field, and parents and educators continue to learn through experience, reflection, and adaptation. Regularly reviewing strategies and observing their impact on children is essential. If an approach is not producing the intended outcome, or if the family feels trapped in a repeating cycle, the strategy should be refined. At this point, learning from specialist advice and successful practices becomes particularly valuable. The central priority remains the wellness of children and finding the most effective ways to guide and support them in an increasingly complex digital world.

Toward a Balanced and Digitally Safe Generation

In closing this chapter, and after reviewing the concepts, guidance, and real stories presented throughout, the overall picture becomes clear. Digital parenting is essential for building a balanced and digitally safe generation in the modern era. Children's digital wellness is not a luxury or a secondary concern. It is a vital educational and life priority that directly affects children's psychological, physical, and social safety. The online world has become an extension of the real world, and separating the two is no longer as feasible as it once was. Preparing a child to become a responsible and ethical digital citizen is therefore a natural extension of raising a responsible citizen in society.

Several key themes have emerged and can be summarized as follows:

Children today grow up immersed in digital environments from their earliest years. This reality creates a dual responsibility, protection and empowerment. Protection involves safeguarding children from risks such as exploitation, cyberbullying, and harmful content. Empowerment, on the other hand, means teaching children how to benefit from the opportunities technology offers for learning, creativity, and growth.

Effective digital parenting rests on multiple pillars, including cybersecurity awareness, information literacy, ethical digital behavior, and healthy balance. When these pillars function together, they form a strong protective framework that supports the child over time, until the child gradually becomes the primary guardian of their own digital behavior.

Each age group has distinct digital needs. There is no single approach that suits every child. A toddler is not the same as a teenager, and each developmental stage requires different strategies. The key lies in tailoring guidance to the child's age and level of maturity, while remaining flexible and willing to adjust rules and expectations as the child grows.

Household rules are not designed to restrict children unnecessarily. Their purpose is to create a safer and more structured digital environment within the home. However, rules become truly effective only when they are supported by a warm and open relationship with children. Dialogue and trust allow guidance to be rooted in care rather than control.

Digital content can act as either a teacher or an enemy. Our responsibility is to choose the teacher and avoid the enemy. A wide range of educational and beneficial applications and games is available today. As digital guides for our children, we can help them access enriching and safe online experiences while steering them away from harmful influences.

Unguided technology use can carry a high cost for children's psychological and physical health. The consequences of neglect in this area may be serious, including emotional distress, isolation, digital addiction, and, in some cases, tragic outcomes. By contrast, investing in awareness and guidance yields important benefits, a happier child, stronger self-confidence, healthier development, and the ability to benefit from technology without becoming controlled by it.

The real stories discussed throughout this chapter served as an honest mirror, reflecting both positive and negative outcomes. Success stories demonstrated that digital interest can be transformed into achievement and meaningful contribution. Risk stories revealed how a small lapse can lead to significant harm, and how awareness and timely action can prevent the worst outcomes and protect children effectively.

Throughout this journey, we focused on teaching children how to protect themselves and how to become the heroes of their own digital story. Just as we equip them with essential life skills, such as basic first aid or crossing the street safely, we must also equip them with digital survival skills, critical thinking, respect for digital privacy, responding to harm, and seeking help when needed. These skills will remain with them and serve as a shield wherever they navigate the online world.

Responsibility does not rest on parents alone. A broader ecosystem also plays a vital role. Aware schools, committed teachers, proactive governments, and responsible technology companies are all working toward the same goal, making the internet a safer and more supportive space for children. Each link in this chain reinforces the others. For this reason, families should make use of every available resource, school programs, awareness initiatives, and parental control tools, and direct them toward children's best interests.

This chapter has shown that no parent is free from making mistakes. What matters most is the willingness to learn, reflect, and adjust. Common mistakes in managing children's technology use can be addressed once they are recognized and acknowledged. The guiding principle throughout has been, and will remain, balance, neither excess nor neglect, but thoughtful moderation.

Building a balanced and digitally safe generation requires patience and continuous effort, but it is an investment worth making. This generation will lead tomorrow, and tomorrow will be even more digital than today. If we prepare children now with awareness, values, and resilience, we can feel confident that, God willing, they will navigate the future responsibly. Just as we care for their physical health, we must also nurture their minds and character through digital knowledge and responsibility. Only then can we benefit fully from technology without allowing its risks to harm the innocence of childhood.

In conclusion, digital wellness is inseparable from wellness in modern life. Our children are a trust placed in our hands, and they deserve our best effort to protect them and support their happiness. Let us walk with them, hand in hand, through the digital world, guiding them when guidance is needed and giving them space to explore with confidence when they are ready. With knowledge, care, and trust in God, we can raise children who are strong in the digital space just as they are strong in real life, and this is the true achievement we seek.

Chapter Seven:

"Digital Wellness in the Workplace"

Technology as a Double-Edged Tool at Work

Technology has brought a profound transformation to the modern workplace. With the spread of smart devices and constant communication through email and messaging applications, employee productivity is now more closely tied to digital tools than at any previous time. For many professionals, a workday rarely begins or ends without a screen.

However, continuous connectivity comes at a cost. What begins as convenience can gradually turn into a digital burden that affects both mental and physical health. In recent years, organizations and researchers have increasingly recognized the importance of digital wellness in the workplace. Digital wellness refers to achieving a conscious and healthy balance in our relationship with technology during working hours and beyond. It does not mean rejecting technology or reducing innovation. Rather, it means using digital tools thoughtfully, in ways that protect productivity while also improving quality of life.

The growing reliance on virtual communication, particularly after the COVID-19 pandemic, has accelerated attention to this issue. Remote work, hybrid models, and constant online collaboration have reshaped how people experience their jobs. As a result, companies and governments have begun to treat digital wellness as a strategic priority because it directly influences employee performance, engagement, and health. For example, the United Arab Emirates established a national council for digital quality of life in

2020, aiming to support balance between digital life and real life and to promote positive and responsible use of technology.

In this chapter, we explore the concept of digital wellness in the workplace, identify common patterns of digital imbalance, and examine their impact on health and productivity. We also present practical strategies and successful local and international models that demonstrate how organizations can build healthier digital work environments.

What Is Digital Wellness in the Workplace Context?

Digital wellness, sometimes referred to as digital wellness, in the workplace means developing a balanced and healthy relationship with technology at work. It is the ability of both employees and organizations to use modern digital tools in a thoughtful and moderate way that supports productivity without sacrificing mental or physical health. This includes managing screen time, controlling the flow of digital communication, and setting clear boundaries for device use during work and after working hours. At its core, digital wellness aims to establish healthy digital habits that reduce technostress and fatigue while supporting a more balanced daily routine.

Although many roles and tasks now depend heavily on the internet and smart devices, technology should not dominate professional life in ways that create harmful outcomes. Digital wellness seeks to align two essential goals: using technology to improve efficiency and performance, while also protecting employee health and reducing stress. Recent studies have suggested links between excessive technology use and mental health concerns such as distraction, anxiety, and reduced concentration. Prolonged digital overload has also been associated with burnout and emotional exhaustion. For this reason, human resources and information technology specialists increasingly advocate for workplace policies that support digital wellness and limit negative effects.

Organizations that adopt digital wellness practices often report tangible benefits. These include stronger employee engagement, higher productivity, improved job satisfaction, and better overall physical and mental health among staff. In practical terms, digital wellness has become an essential component of modern workplace wellness. It provides the balance needed to ensure that technology continues to serve productivity rather than becoming a constant source of pressure and strain.

Signs of Unbalanced Technology Use at Work

Despite the many advantage's technology offers, unbalanced use in the workplace can lead to serious challenges. The following patterns commonly indicate digital imbalance during the workday and highlight how excessive technology use can affect employees' mental and physical wellness.

Digital Meeting Fatigue

Virtual meetings conducted through video platforms have become a central feature of the modern workday, particularly with the rise of remote and hybrid work. Yet when video meetings become excessive, they often contribute to what is widely known as virtual meeting fatigue, or Zoom fatigue. Many employees recognize this feeling as deep exhaustion at the end of a day filled with consecutive video calls.

Workplace experts explain this fatigue as the result of prolonged, high-intensity focus on a screen while simultaneously processing conversation, facial expressions, and visual cues. Participants may also feel pressure to maintain constant camera presence to signal attentiveness, which can create psychological strain and self-monitoring anxiety. In addition, back-to-back meetings leave little room for mental recovery, especially when there is no buffer time between calls.

Common signs of digital meeting fatigue include mental overload, reduced concentration after multiple video meetings, and, in some cases, sleepiness or headaches caused by extended screen exposure. The impact is not only psychological. Physical symptoms such as eye strain, neck tension, and back pain from prolonged sitting are also frequently reported.

In short, when virtual meetings are scheduled without limits or purpose, they shift from being a useful collaboration tool to a source of exhaustion that gradually undermines productivity.

Email Overload and Instant Messaging Pressure

Email and professional messaging applications are essential communication tools in modern organizations. Yet when emails accumulate endlessly, notifications arrive nonstop, and chat messages demand constant attention, employees can quickly experience technostress. Many workers begin their day facing a crowded inbox and feel compelled to respond immediately. Throughout the day, instant messages continue to arrive, keeping attention fragmented and preventing sustained focus.

Research has shown that knowledge workers spend a substantial portion of their workweek on communication alone, including emails, messages, meetings, and coordination. As a result, time available for deep, focused work becomes limited. When reading, writing, and responding to messages consume several hours each day, and meetings are added on top of that, protecting uninterrupted time becomes increasingly difficult. It is therefore unsurprising that many employees report communication overload as a key contributor to burnout.

One common outcome is digital anxiety, the persistent feeling that every message requires an immediate response to avoid appearing inattentive or unprofessional. This creates a continuous state of anticipation, where employees remain mentally alert for the next notification. The fear of missing important information, often described as FOMO, or fear of missing out, further intensifies attachment to devices and heightens anxiety levels.

Another consequence of excessive communication is a decline in communication quality. When too many platforms and messages are used simultaneously, misunderstandings become more frequent and important details can be overlooked. Instead of clarity, digital noise increases complexity and pressure, forcing employees to spend additional energy interpreting instructions and priorities.

In summary, when organizations fail to set clear boundaries around communication volume and responsiveness, technology can shift from an enabling tool into a source of distraction and stress. Over time, this drains employee energy, weakens concentration, and undermines both wellness and performance.

The Blurring of Boundaries Between Work and Personal Life

One of the most serious consequences of digital transformation in the workplace is the gradual erosion of clear boundaries between working time and personal rest. In the past, when an employee left the office, work responsibilities typically paused until the next working day. In today's environment of smartphones and constant internet access, work messages often follow employees into their homes. When boundaries fade, employees may feel as though they are working continuously, always connected and always on call.

This challenge is especially visible in remote and hybrid work arrangements, where professional and personal spaces increasingly overlap. An employee may respond to an urgent email during a family dinner or continue completing unfinished tasks late at night, sacrificing rest and sleep. Professional commentary has increasingly highlighted concerns about work–life balance as digital communication expands and becomes more persistent. These pressures intensified during the COVID-19 period, when large numbers of employees worked from home and the separation between working hours and personal time became even less defined. Digital collaboration increased rapidly, and many employees faced a steady stream of emails, back-to-back video meetings, and continuous chat discussions, often with few genuine breaks during the day.

The likely outcome is exhaustion and the sense that work has expanded to occupy nearly every aspect of life. The loss of boundaries contributes directly to increased psychological strain and can lead to digital burnout, a form of exhaustion linked to constant digital demands. Digital burnout may involve feeling overwhelmed by ongoing tasks, struggling to disconnect from work, and experiencing reduced satisfaction or a weakened sense of professional achievement.

These effects often extend beyond the workplace and into family and social life. People close to the employee may feel that the individual is frequently distracted or emotionally unavailable, which can strain personal relationships. From a health perspective, working beyond normal hours and failing to allow adequate recovery time may contribute to sleep disturbances, persistent physical fatigue, and other stress-related conditions. Health guidance commonly describes burnout as a work-related condition associated with chronic stress that has not been managed effectively. It is typically reflected in exhaustion, increased negative feelings toward work, and reduced professional effectiveness. These outcomes are more likely to emerge or worsen when employees lose the ability to disengage from digital work demands outside official working hours.

For this reason, some countries, including France, have introduced legal and policy measures often described as the "right to disconnect" after working hours. These initiatives aim to reduce long-term strain and protect employees from expectations of constant availability. In workplaces where such norms are weak or absent, employees may feel obliged to remain reachable at all times, trapping them in a cycle of ongoing pressure and fatigue.

Psychological and Physical Effects of Digital Imbalance

Signs of digital imbalance in workplace technology use affect employee health on multiple levels. Psychologically, continuous technostress can increase anxiety and tension and may result in persistent mental fatigue. Many employees describe ending the workday feeling mentally drained due to the sheer volume of alerts, messages, and information they have processed. They may also experience a reduced sense of control, fragmented attention, and difficulty concentrating on a single task because of frequent switching between digital demands. Over time, these patterns can develop into burnout, which may appear as low motivation, emotional exhaustion, and a sense of disconnection from colleagues.

Digital pressure can also contribute to broader mental health challenges, including anxiety and depression. Constant notifications and expectations of immediate response keep the nervous system in a heightened state of alert, similar to a continuous fight-or-flight response. This makes it difficult for the body and mind to relax. Occupational health specialists frequently note that heavy screen dependence and ongoing connectivity are associated with higher levels of stress and anxiety.

Physically, the consequences of digital imbalance can be equally serious. Prolonged sitting in front of screens without sufficient movement may contribute to musculoskeletal problems such as neck and back pain, as well as repetitive strain injuries, including carpal tunnel syndrome, caused by extended keyboard and mouse use. Excessive screen exposure, particularly in the evening, can disrupt sleep patterns due to blue light emission, leading to insomnia or poor-quality sleep. Many employees also report frequent headaches and eye strain resulting from long periods of screen focus without breaks. Additional symptoms may include dry eyes and blurred vision.

Reduced physical activity linked to heavy digital engagement may also contribute to long-term health risks, including weight gain and cardiovascular disease, due to a more sedentary lifestyle. These effects highlight that digital wellness is not limited to psychological comfort but also encompasses physical health and safety.

A central paradox emerges in this context. Tools designed to increase efficiency and productivity can produce the opposite outcome when used without balance. Digitally exhausted employees often demonstrate lower productivity and creativity, make more errors, and show weaker decision-making due to fatigue and reduced concentration. In some cases, prolonged digital strain can even lead skilled employees to leave their jobs in search of healthier and less stressful environments, increasing staff turnover.

One recent survey found that 64 percent of employees reported lacking the time or energy needed to complete their work effectively because of ineffective meetings and frequent interruptions. This pattern also significantly reduced their ability to engage in creative and strategic thinking. The result is a negative cycle in which declining health and declining productivity reinforce each other. A digitally exhausted employee performs less effectively, and weaker performance creates additional stress, further deepening exhaustion.

For this reason, leading organizations increasingly recognize that investing in digital wellness is also an investment in organizational success. In the following sections, we will explore practical ways to break this cycle by adopting effective strategies that restore balance and strengthen digital wellness in the workplace.

Practical Strategies for Achieving Digital Wellness at Work

Fortunately, both organizations and individuals can take practical steps to reduce digital strain and strengthen digital wellness in the workplace. Achieving this balance requires a combination of institutional policies and individual habits that help restore structure and rhythm to the workday. The following strategies have been widely adopted as effective tools for supporting healthier and more sustainable digital work practices.

Scheduling Periods Without Meetings or Constant Communication

It can be highly effective to designate specific periods during the week that are free from digital meetings and instant messaging, allowing employees to focus on deep work without constant interruptions. Some organizations adopt practices such as a "no-meeting Wednesday" or scheduled daily focus blocks during which interruptions are intentionally minimized. This strategy directly addresses attention fragmentation and gives employees protected time to complete complex and strategic tasks.

Neuroscience-oriented discussions suggest that dedicating time to uninterrupted work, including periods away from screens, can enhance problem-solving ability and overall performance. Even simple actions, such as turning off notifications or placing a mobile phone out of reach during focus periods, can produce noticeable improvements in productivity and the quality of thinking.

Adopting a "Right to Disconnect" Outside Working Hours

A core principle of digital wellness is the establishment of clear temporal boundaries for work communication beyond official working hours. Organizations can support this by formally recognizing employees' right to disconnect at the end of the workday, meaning they are not expected to respond to emails or messages after hours. This type of policy has already been introduced in several international workplaces. One frequently cited approach involves limiting or delaying non-urgent work emails outside normal working time in order to reduce stress and support recovery without undermining productivity.

Despite these examples, surveys suggest that many organizations still lack clear policies to prevent after-hours communication. This highlights the need to normalize disconnection as an accepted and protected part of workplace culture. Leadership plays a critical role in reinforcing this norm by sending a clear message: employees are entitled to personal time without guilt or penalty. Organizations can also support this in practical ways, such as delaying the delivery of messages sent late in the day or encouraging email scheduling so communications arrive the following morning. These measures allow employees to gain the mental rest required for recovery, supporting stronger energy, focus, and performance during working hours.

Establishing Responsible Use Policies and Building Team Awareness

Introducing policies alone is rarely sufficient unless they are supported by shared understanding and translated into daily practice. Organizations should therefore educate employees, through training sessions and workshops, on effective methods for managing digital time and reducing technostress. This may include guidance on organizing email inboxes, adjusting application settings to limit disruptive notifications, and following clear communication norms. For example, teams can agree on when email is more appropriate than instant messaging, who genuinely needs to be included in communications, and what response times are considered reasonable.

Such awareness helps reduce unnecessary message volume and encourages more intentional and respectful communication. Teams may also benefit from developing an internal charter for responsible technology use. For instance, they may agree that no responses are expected during weekends or that messages should not be sent after a specified evening time. Management guidance often recommends that leaders support digital wellness by enabling employees to complete their work within normal hours, setting clear expectations for virtual behavior—such as limiting the number of meetings per day and defining maximum meeting lengths—and actively encouraging breaks and leave. When employees perceive that leadership genuinely values their digital health, adherence to these practices becomes more natural. Human resources therefore plays a central role in embedding digital wellness into workplace policy, training, and long-term organizational culture.

Designing a Work Environment That Supports Focus and Health

Organizations can also reduce digital strain through relatively simple spatial and organizational adjustments. For example, some workplaces have introduced quiet areas that are free from electronic devices, where phones and computers are deliberately excluded. These spaces allow employees to work or rest without digital distraction.

In the United Arab Emirates, the Dubai Electricity and Water Authority has established dedicated focus rooms where employees can complete tasks with fewer interruptions and hold meetings without mobile phones. In several European contexts, the concept of digital rest spaces has also gained traction, referring to screen-free areas designed to restore mental clarity. This model has been adopted in offices across major cities.

Beyond workspaces, organizations can redesign break areas to encourage social interaction and restorative activities away from screens. This may involve providing books, magazines, or simple games instead of placing televisions and digital displays in every area.

Workplace studies suggest that reintroducing screen-free breaks can support sustained concentration and mental recovery.

Promoting physical activity during the workday is equally important. Providing space for walking, offering light exercise options, or organizing group health initiatives such as daily step challenges can reduce prolonged sitting and support overall wellness. Visual health should also be considered, for example by promoting the 20-20-20 rule, which encourages employees to rest their eyes every twenty minutes by looking at something twenty feet away for twenty seconds. Organizations may also consider screen filters or blue-light-filtering glasses for employees who experience frequent eye strain. While these measures may appear modest, they send a clear signal that digital health is an integral part of a healthy work environment.

Using Technology to Support Digital Wellness

Although it may appear paradoxical, technology itself can play a valuable role in addressing technology-related strain. Many modern tools and platform features are specifically designed to support digital wellness. For instance, organizations can activate Do Not Disturb settings within operating systems and collaboration platforms during focus periods or outside working hours to block non-essential notifications.

Even simple experiments, such as enabling Do Not Disturb mode for a twenty-four-hour period, have been reported to improve productivity and reduce perceived stress. Organizations can also encourage employees to monitor their screen use through digital wellness applications that provide alerts when time limits are exceeded or remind users to take breaks. Many smartphones now include built-in dashboards that display daily application usage and allow users to set usage limits. This increased self-awareness can help employees adjust habits and regain control over their attention.

Automation can further reduce digital burden by streamlining routine tasks. Tools that schedule messages, manage approvals, or automate repetitive administrative processes can minimize unnecessary follow-up communication and reduce distraction. Workplace reports suggest that automation of routine digital tasks can free up time and mental space for more creative and strategic work.

Finally, organizations that rely on multiple platforms and applications should regularly evaluate how effectively these tools integrate. An excess of disconnected systems can create tool fatigue rather than improving efficiency. In many cases, adopting more integrated platforms can reduce cognitive load and limit the need for employees to constantly switch between applications, thereby supporting a more balanced and sustainable digital work experience.

Leadership Role Modeling and Commitment to Digital Wellness

No policy or strategy can succeed without genuine commitment from organizational leadership. Managers and executives play a crucial role in modeling healthy digital habits. When leaders send emails in the middle of the night or schedule meetings at all hours, employees are unlikely to feel safe setting boundaries. By contrast, when leaders visibly respect personal time, such as avoiding after-hours communication or openly sharing their own efforts to disconnect during weekends, the workplace culture often begins to shift in a more balanced direction.

Workplace experience frequently shows that senior leadership commitment is one of the fastest drivers of cultural change. In some organizations, digital discipline is even treated as a performance expectation. One manager at an investment company in the United Arab Emirates was quoted as saying that employees are evaluated partly on their digital discipline, and that seeing a mobile phone used during meetings signals limited control over attention. While not every workplace would adopt such a strict stance, the underlying message reinforces the value of focus, presence, and higher-quality work.

Leadership can further strengthen digital wellness by launching internal initiatives that actively challenge unhealthy norms. Examples include setting a temporary rule to avoid meetings after mid-afternoon for a week, or introducing a day without email to encourage alternative forms of communication. Involving employees in shaping solutions, through working groups, pilot programs, or regular surveys, can also be highly effective, as employees often have practical insights into reducing digital strain in their daily tasks.

In summary, when employees observe that leadership treats digital wellness as a genuine priority rather than a discretionary benefit, they are far more likely to engage with, support, and sustain the behavioral changes required.

Examples of Institutions That Apply a Digital Wellness Culture

Several real-world examples from the Gulf region and beyond illustrate how organizations are actively building a culture of digital wellness. These cases demonstrate that relatively simple shifts, when thoughtfully designed, can produce meaningful improvements in employee health and productivity.

Government Institutions in the United Arab Emirates

The United Arab Emirates is widely regarded as a regional leader in promoting digital wellness at a national level. In addition to establishing a national body focused on digital quality of life to guide policy and raise public awareness, the government has implemented practical measures to improve work–life balance. In 2022, the public sector introduced a four-and-a-half-day workweek, running from Monday to mid-Friday. The initiative aimed to enhance employee quality of life while maintaining institutional performance, and early assessments following implementation reported improvements in both employee wellness and operational efficiency.

In 2024, the Government of Dubai launched a pilot initiative titled "Our Summer Is Flexible," which tested a four-day workweek during the summer across selected government entities to evaluate its effects on productivity and job satisfaction. In parallel, some public organizations, including the Dubai Electricity and Water Authority, have introduced dedicated focus rooms designed to reduce digital distraction. These spaces allow employees to work with fewer interruptions and support more disciplined, phone-free meetings. Together, these examples highlight that digital wellness has moved beyond isolated initiatives and is now embedded within a broader governmental vision for quality of life.

Leading Companies in the Gulf Region

Across the private sector in the Gulf, many organizations have begun adopting practices that support digital wellness. A notable example is Lusidia, a Saudi technology startup based in Riyadh, which announced in 2024 that it would introduce a four-day workweek. The company framed the decision as a step toward a more modern and sustainable workplace, emphasizing that service quality would not decline. Leadership also highlighted that additional rest time could improve motivation, support personal development, and allow employees to return to work with greater focus and commitment. This approach aligns closely with findings reported in international research.

One widely cited field trial in the United Kingdom, involving sixty-one companies and approximately 2,900 employees, found that a four-day workweek was associated with improvements in mental and physical health, alongside reduced levels of stress, anxiety, and exhaustion. Many participants also reported better balance between work, family, and social life, while revenue levels were maintained in a significant number of participating organizations. These results suggest that a mutually beneficial model is achievable, in which employee wellness improves without undermining organizational performance.

A similar example often referenced is Microsoft Japan, which reported in 2019 that a trial offering a four-day workweek led to a substantial increase in productivity, alongside reductions in electricity use and paper consumption. Within the Gulf, other organizations have adopted complementary measures such as flexible working hours, structured remote work policies, and internal communication guidelines that discourage non-essential messages during rest periods. Although many of these practices are still evolving, they reflect a growing recognition that long-term organizational sustainability depends not only on financial outcomes, but also on preserving the energy and wellness of human capital.

Inspiring Global Examples

Internationally, several countries and organizations provide strong examples of how digital wellness can be embedded into workplace culture. In Europe, France introduced a legally recognized "right to disconnect," allowing employees to ignore work-related digital communication outside working hours without professional consequences.

In Germany, major companies such as Mercedes-Benz have offered employees the option to automatically delete incoming emails during vacation periods, preventing the stress associated with returning to an overwhelming inbox. Volkswagen took an early step in a similar direction by restricting the delivery of work emails to employees outside official working hours.

In the United Kingdom and Scandinavian countries, many offices have established device-free spaces where employees can step away from screens and regain mental calm during the day. Even in high-pressure financial environments, some firms have experimented with initiatives such as "screen-free Fridays," dedicating time to idea generation and collaboration that minimizes electronic device use. In several cases, these measures have been linked to reduced strain and improved performance, encouraging offices in cities such as London and Stockholm to adopt quiet work zones with limited or no technology.

Iceland provides another compelling example through national trials of shorter working weeks, typically around thirty-five hours, which have been associated with stable or improved productivity, higher employee wellness, and lower levels of stress and burnout. The Netherlands similarly demonstrates how flexible work arrangements and shorter average working hours can coexist with strong economic performance. Together, these cases suggest that productivity and quality of life are not mutually exclusive.

The overarching lesson from these global examples is clear. Organizations that deliberately rethink how work is structured—by reducing excessive hours, creating space for digital recovery, or giving employees greater control over time and communication—often cultivate a healthier, more engaged workforce without sacrificing performance. These insights can be adapted to local contexts to strengthen digital wellness as a lasting component of institutional culture.

Statistics and Studies on the Impact of Technology on Productivity and Burnout

Global data highlights the growing seriousness of digital strain and its measurable impact on workplace performance. The following statistics are frequently cited in research and professional reports to illustrate the scale and urgency of the issue.

In several surveys, approximately 60 percent of employees reported that excessive digital communication is a major contributor to burnout. This finding suggests that many workers experience constant messages, alerts, and notifications as mentally exhausting rather than supportive.

Knowledge workers are reported to spend as much as 88 percent of their workweek engaged in communication activities, including email, meetings, and coordination across digital platforms. This leaves very limited time for uninterrupted, focused work. On average, some estimates indicate that employees may spend around twenty hours per week using digital communication tools alone.

A global Microsoft survey published in 2022 revealed that 64 percent of workers felt they lacked sufficient time or energy to complete their tasks due to the volume of meetings and inefficient communication practices. The same survey also found that 64 percent of respondents believed constant interruptions prevented them from engaging in creative or strategic thinking, highlighting a direct link between digital overload and reduced innovation.

One European study reported a sharp rise in the proportion of employees experiencing digital strain, increasing from 8.5 percent in 2020 to 20.2 percent in 2022. This substantial increase over a short period suggests that digital pressure intensified during and after the pandemic, as remote and hybrid work became more common.

In addition, some reports indicate that more than 70 percent of leaders in human resources and internal communications have observed rising levels of employee burnout and directly associate this trend with excessive digital load in the workplace. This alignment among organizational decision makers reinforces the need for deliberate and structured intervention.

Despite growing awareness, a significant gap remains between recognition and action. Around 74 percent of human resources leaders in certain surveys acknowledged that their organizations had not implemented measures to reduce excessive internal communication. Approximately 70 percent also reported the absence of policies limiting contact outside official working hours. This disconnect underscores the need for faster and more decisive organizational responses.

Taken together, these figures convey a consistent and concerning message. Overuse of workplace technology can drain employee energy and weaken productivity, yet many organizations have not fully responded with practical solutions. Drawing on recent data and research can help persuade decision makers that fostering a culture of digital wellness is no longer optional. It is a necessity for protecting both employee health and organizational performance.

Recommendations for Employers and Decision Makers

Embedding digital wellness into organizational culture requires visible leadership commitment and structured actions that normalize healthy digital practices. The following recommendations outline practical steps employers and decision makers can take to strengthen digital wellness and integrate it into everyday workplace operations.

- **Establish a formal framework and official digital wellness policies:**

Organizations should begin by adopting a clear policy that defines healthy technology use. This framework may include a digital wellness charter that outlines employee rights and responsibilities related to after-hours communication, expected response times, and appropriate meeting frequency.

A policy of this nature sends a strong signal that leadership is committed to supporting digital balance rather than leaving boundaries to individual negotiation. Digital wellness can also be embedded within human resources policies related to occupational health and safety. For example, the policy may specify that employees are not expected to respond to messages after seven in the evening, or it may place limits on the number of daily meetings and their duration, such as a maximum of forty-five minutes. When these expectations are formalized, they become shared organizational standards rather than personal preferences.

- **Communicate digital expectations clearly and consistently:**

Leaders and managers should address digital wellness openly and regularly with their teams. Employees need to hear clearly that completing work within official working hours is the norm, and that overtime should remain an exception rather than an unspoken expectation. It is equally important to communicate that not responding during evenings, weekends, or leave periods is acceptable and encouraged for health and recovery.

Clear verbal and written guidance can significantly reduce the invisible pressure many employees experience. Organizations may also hold listening sessions to identify the main sources of digital strain and develop targeted responses based on employee feedback. Appointing a digital wellness champion within each team or department can further support implementation by reinforcing agreed practices and acting as a bridge between staff and leadership.

- **Integrate digital wellness into performance assessment and organizational culture:**

At a broader level, digital wellness should be embedded into organizational values and everyday culture. Just as organizations promote values such as teamwork, integrity, and innovation, they can also promote respect for time, boundaries, and health in digital work.

Organizations can incorporate digital balance indicators into employee satisfaction surveys. If a large proportion of staff report difficulty disconnecting, this should be treated as an early warning sign requiring corrective action. Digital wellness can also be reinforced through performance evaluation by recognizing positive behaviors, such as effective collaboration without excessive meetings and consistent respect for after-hours communication guidelines. When employees see that healthy digital discipline is valued, they are more likely to adopt and sustain it.

- **Provide training and resources for employees:**

Many employees lack practical skills for managing technology in ways that support focus rather than drain attention. Employers can address this by offering targeted training and resources. Workshops on digital time management, focus techniques, and prioritization in high-distraction environments can be particularly effective. Strategies such as the Pomodoro technique can help employees structure work and rest periods more intentionally.

Employees should also be trained to use available platform features effectively, including notification controls, status indicators to signal focus time, and simple workflow tools that reduce unnecessary interruptions. Some organizations provide access to applications that support mindfulness, task management, or healthy habit tracking. Although modest in cost, these resources signal genuine concern for employee wellness and provide practical tools for managing fatigue.

Promoting a culture of effective breaks is equally important. Employers can share guidance on desk stretching, encourage regular movement, and remind employees to rest their eyes by looking away from screens periodically. When these practices are consistently reinforced, they can become part of everyday work routines.

- **Redesign work structures toward flexibility:**

Employers are well positioned to rethink traditional work structures to better reflect the realities of the digital era. Flexible working hours can allow employees to take longer breaks when needed and adjust schedules to support energy and focus. Where feasible, organizations may also explore reduced working models, such as a four-day workweek.

Flexibility enables employees to organize their lives in ways that reduce stress and enhance performance during working hours. A critical component of this shift is focusing on outcomes rather than constant online presence. Leaders should emphasize results and work quality rather than continuous availability. This approach reduces pressure to remain perpetually connected and encourages more intentional time management followed by genuine rest.

As part of work redesign, employers should also review workload distribution. Digital strain is often a symptom of unrealistic expectations that force employees to work beyond normal hours. Balanced workloads, combined with adequate human and technical resources, are essential to ensure tasks can be completed within official working time and to prevent long-term burnout.

- **Lead Change from the Top:**

Finally, any lasting cultural shift depends first and foremost on senior leadership. Decision makers must be the first to model digital wellness behaviors in practice, not only in policy. Chief executives and senior managers can lead by example by launching internal initiatives on digital wellness and by sharing personal reflections on how they have adjusted their own working habits to achieve healthier balance.

Visible, consistent support from leadership gives credibility and momentum to any digital wellness initiative. When employees see leaders respecting boundaries and prioritizing balance, they are more likely to feel safe doing the same.

Leaders should also monitor progress and define clear, measurable goals. For example, an organization may aim to reduce the average number of emails per employee by twenty percent within a quarter, or increase the proportion of employees reporting a healthy work–life balance by fifteen percent in the annual survey. Setting concrete targets signals that digital wellness is tied to outcomes, not just good intentions. When progress is achieved, it is important to recognize it, whether by acknowledging a department that successfully reduced unnecessary meetings or by highlighting improvements in satisfaction and reductions in burnout through internal communications. A positive and reinforcing culture helps sustain change and embed it within organizational identity.

As one workplace experience specialist observed, "By prioritizing wellness alongside productivity, organizations can build a happier, healthier, and more engaged workforce." This reflects the direction many institutions must pursue in the coming years.

Quality of Life Depends on Digital Wellness

In conclusion, digital wellness at work is not an optional luxury. It is a core requirement for the long-term success of both individuals and institutions. We have entered an era in which the ability to disconnect from technology at appropriate times can be as valuable as the ability to master it. As the examples discussed show, organizations that adopt a balanced approach to technology gain tangible benefits in productivity, commitment, and employee retention. When employees feel rested and satisfied, they make fewer mistakes and demonstrate stronger creativity and focus.

At the individual level, healthier digital habits at work directly enhance overall quality of life. An employee who can close their computer in the evening and dedicate time to family, rest, and personal interests is more likely to feel fulfilled and resilient. This sense of balance, in turn, increases readiness to perform well professionally. It creates a reinforcing cycle in which healthy work supports a healthy life, and a healthy life supports better work.

More broadly, quality of life is increasingly shaped by our ability to maintain balance between the digital world and real life. In an environment driven by constant connectivity and fear of missing out, the advantage often belongs to those who have the discipline and confidence to step away from digital noise when needed. The solution is not simply more applications or new technical fixes, but a cultural shift that ensures technology remains in service of people, rather than people being driven by technology.

The future of work will favor those who know when to pause, lift their attention to the bigger picture, and engage with others thoughtfully and with clarity. Those who are absorbed in constant digital flow without boundaries may lose focus, creativity, and wellness without realizing it. Some of the most meaningful innovations of the future may arise not from working longer hours, but from working more intelligently and reducing unnecessary effort. Building a culture of digital wellness is a foundational step toward achieving that future.

With strong leadership commitment and growing employee awareness, workplaces can evolve into more humane and balanced environments, where productivity develops alongside happiness and health. Only then can technological progress remain a true asset that enriches professional and personal life, rather than a burden that drains energy and wellness.

Chapter Eight:
"Practical Guidelines and Applications for Digital Wellness"

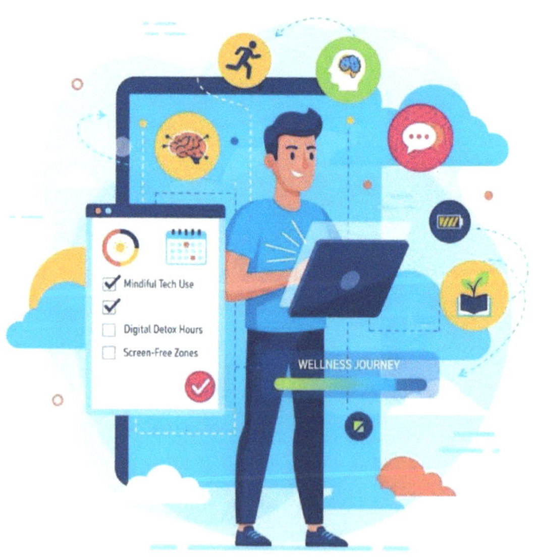

In the era of smartphones and constant connectivity, achieving digital wellness has become a shared challenge for everyone, from parents and teachers to employees and young people. Excessive immersion in the digital world without clear boundaries can lead to fatigue, stress, and weaker social connection. By contrast, adopting healthy and intentional practices in the use of technology supports a more balanced relationship between our online lives and our lives beyond the screen.

In this chapter, we present practical strategies for strengthening digital wellness in everyday life, along with models for organizing time across work, family, rest, and technology. We also include real-world examples and success stories, as well as engaging challenges designed to motivate reduced and more mindful use. In addition, we discuss how these practices influence mental health, relationships, and productivity, and we provide guidance for developing an effective personal plan. The overall goal is to inspire readers to turn technology into a tool that supports their goals rather than an obstacle. When we manage our digital lives with awareness, we create stronger opportunities for personal growth and family wellness.

Practical Strategies to Improve Digital Wellness

Improving digital wellness requires clear and realistic steps that can be applied consistently. The following strategies can be integrated into daily life to support a healthier and more balanced relationship with technology.

1.   Disable Non-Essential Notifications:

Constant alerts from smartphones and applications are among the biggest sources of distraction and stress. It is helpful to turn off, or at least reduce, notifications from non-essential apps and to keep the phone silent during periods of focus or rest. Many studies link high notification volume with increased anxiety, so limiting alerts to what is truly necessary, such as calls or urgent messages, can restore calm and concentration.

You can also use features such as Do Not Disturb or Focus Mode, which mute non-essential notifications during selected times while allowing only important contacts to reach you. This prevents the phone from pulling your attention away from the present moment, whether you are studying, resting, or spending time with family.

2.   Schedule Regular Digital Breaks:

Including screen-free breaks throughout the day is essential. Long, uninterrupted screen time strains both the eyes and the mind, and it can also contribute to physical discomfort, including posture-related pain often referred to as "tech neck." For this reason, it is recommended to stand up regularly during the workday, walk briefly, stretch, or do light exercises away from screens.

Experts often suggest short digital breaks across the day, such as taking five to ten minutes each hour to look away from the screen or engage in a non-digital activity. Even very brief breaks can noticeably reduce stress and mental fatigue. One study found that dedicating just thirty minutes a day to disconnecting from the phone can improve life satisfaction and reduce screen-related pressure.

Turning these pauses into a habit, such as enjoying a coffee break without a phone or taking a short midday walk outdoors, helps restore energy and supports a return to screens with better focus and clarity.

3.   Manage your Screen Time Deliberately:

Screen time management lies at the heart of digital wellness. Set daily or weekly limits for device use and for the applications that consume the most time. For example, you might allocate one hour in the evening for social media instead of leaving usage open-ended. Most smartphones now include screen-time tools that track how long you spend on each app, which increases awareness of your habits.

After reviewing your usage patterns, set limits for the most time-consuming apps, such as social media platforms or games. When the daily limit is reached, the phone can notify you that you have crossed the boundary. These small adjustments train you to manage digital time with intention rather than impulse.

It is also helpful to assign specific windows for certain tasks. For instance, you might check email only twice a day instead of monitoring it constantly, or read the news at a fixed time in the morning rather than drifting into it repeatedly. This structure reduces wasted hours spent moving aimlessly between applications.

A core principle of digital wellness is simple: technology should work for you, not control your time or attention.

4.   Define Purposeful Phone and App Use:

Being selective about phone use is highly recommended. Ask yourself: what are the essential tasks I truly need my phone for, and which apps tend to steal my time without offering real benefit? Based on this reflection, classify your apps into useful, essential tools and those that mainly serve as entertainment or time-fillers. Try to delete, or at least reduce, apps that do not add genuine value to your life.

For example, if you notice hours spent scrolling through content that does not matter to you, consider removing that app or replacing it with an educational or wellness-focused alternative. Set clear goals for phone use, such as staying connected with loved ones, learning something new daily, or using your device to plan and organize tasks. You can also create personal rules, such as no phone use during meetings or study time, and no random browsing before sleep.

Experts also advise establishing device-free zones at home. Phones can be kept away from the dining table to support real conversation during meals, and left outside the bedroom at night to promote better sleep. These time-based and place-based boundaries help protect the balance between digital and offline life.

Ultimately, digital wellness depends on intentional use. The goal is to engage with technology for clear purposes, work, communication, or learning, rather than slipping into aimless scrolling that simply fills empty moments.

Expert note:

Dubai-based psychologist Amna Hussein advises: "We should monitor our online activity, use our screen time more wisely, set clear boundaries between work and entertainment, and designate device-free areas at home. It also helps to step away from digital toxins for a period, spend time moving, reflecting, and being in nature. This is necessary for our digital wellness."

This advice captures the core message: observe your habits, adjust them thoughtfully, set boundaries, and regularly reclaim time for yourself away from digital noise.

Weekly Schedule Templates for Digital and Life Balance

Planning the week in advance can support a healthier balance between work, family life, rest, and technology use. Below is a suggested weekly structure that can be adapted to individual needs.

1. During Workdays (Sunday to Thursday):

Begin the morning with a calm activity without checking your phone immediately after waking, such as a short mindfulness exercise or breakfast with family. Avoid opening social media first thing in the morning, as starting the day without it strengthens your ability to manage attention later.

During work or study hours, focus on scheduled tasks and take short device-free breaks, for example once every hour. If needed, allocate five minutes every two hours to check your phone instead of remaining distracted throughout the day. Make lunch as screen-free as possible, using the time to talk with colleagues or eat slowly and mindfully.

After the workday ends, separate professional time from personal time. Complete essential tasks, then shut down your work computer and place your work phone out of reach. In the evening, give family and yourself meaningful attention. Sit together, talk, or eat dinner without phones, and agree as a household to keep devices away. Research suggests that even the presence of a phone on the table can reduce the quality of face-to-face conversation.

You may set aside one hour for watching television or a family program together, but avoid using other devices at the same time. Finally, at least one hour before sleep, turn off electronic devices or enable night mode. Light reading or relaxation exercises before bed are healthier alternatives to phone screens, whose blue light can interfere with sleep quality.

Following these habits during the week can improve productivity and concentration while ensuring that technology does not consume personal time.

2.   During the Weekend:

Weekends provide a valuable opportunity to recharge mentally and emotionally, away from both work pressures and constant digital demands. Plan family and social activities to balance the many hours spent on screens during the week. For example, dedicate one day to family activities such as outings, group sports, or visiting relatives.

It can be helpful to declare a "device-free day," or at least a device-free portion of the day, by placing phones in a drawer and focusing on shared experiences. At first, this may feel uncomfortable or boring, but the time can be filled with enjoyable activities such as board games, cooking together, creative projects, or spending time in nature. Many families are surprised by how much conversation, laughter, and connection emerge once phones are removed from the scene.

As for Saturday, or the other day of the weekend, it can be approached in a less restrictive way digitally while still maintaining balance between responsibility and enjoyment. You may allow time for watching television or playing video games in moderation, especially for children, while also planning shared physical or recreational activities. For example, the morning can be dedicated to exercise or a walk in the park; the afternoon can allow free time for each family member to enjoy activities of their choice, including limited screen time; and the evening can be reserved for gathering together as a family to watch a movie or engage in meaningful conversation.

The key is to schedule digital time consciously, even on days off, rather than allowing it to creep in and consume the entire day of rest. If you feel the urge to browse social media over the weekend, try doing so only after completing something beneficial, such as finishing a chapter of a book or visiting friends. Some people also find it helpful to adopt a "no social media weekend" rule, which involves completely avoiding platforms such as Facebook, X, and Instagram from Thursday evening until Sunday morning. This form of digital detox can clear the mind from constant updates and information overload.

3. Allocating Fixed Times for Each Area:

The secret to achieving balance lies in giving every aspect of life its fair share, without excess or neglect. Within any weekly schedule, it is important to include time for work and productivity, family and social relationships, rest and sleep, hobbies or self-care and health, as well as limited and well-planned time for technology. For example, you might decide that every evening after 8 p.m. is device-free family time, or that Saturday morning is dedicated to exercise or a personal hobby. These commitments should be written into your schedule just as you would schedule meetings or work tasks, because honoring them is essential for maintaining quality of life. A week in which time is balanced across diverse activities often ends with a sense of satisfaction and renewal, rather than exhaustion. Try different approaches until you find a routine that fits your lifestyle and supports your family, professional, and personal needs together.

Real-Life Examples: Success Stories in Adopting Digital Wellness

The strongest evidence of the effectiveness of digital wellness guidelines often comes from real-life experiences of individuals and families who have made meaningful changes. The following examples from the Gulf and beyond illustrate how small shifts can lead to positive outcomes.

A day without a phone, Tia's experience (16 years old):

Tia is a teenager whose mother felt that excessive smartphone use was beginning to affect her studies and health. At her mother's request, Tia agreed to stay away from her phone for a full day. Although hesitant at first, she quickly became overwhelmed by feelings of anxiety and emptiness, as she was accustomed to checking her phone every few minutes. Determined to cope, she decided to fill the gap with something else. She went to the home library and picked up a book she had started long ago but never finished. She soon became deeply absorbed in reading, experiencing an unfamiliar sense of psychological relief. Tia explains, "Even though the beginning was hard, I felt a strange calm when I got lost in reading. I remembered how much I loved it in my childhood."

By the end of the day, Tia realized she could manage twenty-four hours without her phone. As a result, she began gradually reducing her daily screen time and committed to setting aside time for reading each day. Although she admits that giving up her phone entirely for a full day still feels challenging, she has become more aware and more motivated to maintain a healthier balance.

Anas and Digital Calm (18 years old):

Anas is another young person who chose to experiment with a phone-free day after feeling intense psychological pressure from constant notifications and nonstop messages. "I felt mentally exhausted," he says. "My friends suggested I try staying away from my phone for a day."

At first, boredom set in, and he felt lost without his device. "I cannot deny it," he admits, "my hand kept searching for my phone automatically." Even his thoughts needed time to adjust to the absence of digital noise. However, as the day progressed, Anas began to feel calmer and more at ease. He used the time to rest and reflect on personal matters he usually ignored while constantly connected. "That day was a chance to reconnect with myself away from the digital chaos," he explains.

This experience led Anas to adopt a new habit: dedicating one day each week to minimal phone use, avoiding non-essential apps and allowing only urgent calls or messages. This weekly practice has become a personal ritual for restoring mental energy and reducing stress.

Tangible Benefits of these Experiences:

The stories of Tia and Anas reflect what scientific research consistently shows about the benefits of reducing screen use. Studies have found that avoiding phone use for even one day can lower stress and anxiety levels. Excessive smartphone use is associated with reduced focus and increased tension, while short periods of digital fasting can enhance psychological comfort and concentration. Research discussed in "Psychology Today" has also noted that stepping away from phones allows the mind to rest from constant stimulation, creating space for hobbies, better sleep, and improved creative thinking.

On a social level, digital breaks often strengthen real-world connections. A phone-free day frequently leads to longer conversations, shared activities, and deeper interaction with family and friends. In simple terms, relationships improve when attention shifts from screens to face-to-face engagement.

A Family Reconnects: A Screen-Free Weekend:

In a real example from abroad, a mother of two children aged eleven and thirteen shared her experience of a screen-free weekend, during which the family disconnected from devices for two full days. The idea emerged after she noticed that screens had begun to dominate family time, even when everyone was together.

Initially, the proposal was met with strong resistance. Her teenage son called it "the worst idea ever." Nevertheless, she encouraged the family to treat it as a short experiment. On Friday evening, they placed all devices in a box out of reach and informed relatives that they would be offline until Sunday.

The first hours felt slow, boredom appeared, and the question arose: "What do we do now?" Gradually, however, the family filled the time with meaningful activities. They went on long walks, played board games, cooked meals together, and resumed a puzzle that had been forgotten on a shelf. One particularly memorable moment was a nature outing without phones, filled with uninterrupted conversation and laughter.

The mother later shared that she learned more about her children's school lives and friendships during that weekend than she had in weeks. By the end of the two days, everyone felt calmer and happier, and no one felt they had missed anything important online. Instead, they gained a shared experience that motivated them to repeat screen-free weekends regularly.

The Right to Disconnect from Work: A Professional Initiative:

Efforts to improve digital wellness extend beyond individuals and families to institutions and governments. In France, for example, a law known as the "Right to Disconnect" was enacted in 2017, granting employees the right to ignore work emails and calls outside official working hours. This legislation followed research showing that constant work-related connectivity leads to stress, insomnia, and strained personal relationships. Today, companies with more than fifty employees are required to negotiate a charter that clearly defines the employee's right to disconnect.

This initiative reflects growing recognition that work-life balance includes a digital dimension, and that constant availability is harmful.

Increasingly, organizations in various countries, including some in the region, are encouraging employees to avoid checking email at night or during vacations in order to protect mental health and sustain long-term productivity.

A Fun Family Challenge: Turning Screen Reduction into Friendly Competition

In another lighthearted example reported by The Wall Street Journal and translated by Al Jazeera, digital health expert Chris Flack suggested a creative way to motivate families to reduce screen use through friendly competition. Flack proposes that if a family enjoys a competitive spirit, it can be transformed into a shared challenge aimed at cutting down digital consumption. For instance, parents may challenge themselves to avoid browsing Instagram for two hours per week, while children commit to reducing their TikTok use by the same amount. Whoever successfully meets the goal earns a reward, such as going out for pizza, with the cost paid by the "losers," for example.

This approach turns screen time management into an engaging game rather than a rule imposed from above. Flack noted that these family challenges foster an atmosphere of mutual support, as family members encourage one another to stay committed in order to win. The result is not only a shared meal, but also greater awareness of how time is spent on devices and a deeper appreciation of real time together.

These real-world examples demonstrate that small changes can lead to meaningful results, whether through an individual experience of a day without a phone, a family initiative such as a screen-free weekend, or institutional policies that promote clearer boundaries between work and constant connectivity. They all point to one central truth: when we control technology instead of allowing it to control us, we reclaim our time and create space for greater happiness and achievement.

Motivational Digital Challenges to Experience Digital Wellness

Reducing dependence on devices can feel intimidating at first, which is why turning the process into a voluntary and enjoyable challenge can be highly effective. The following practical challenges can be tried individually or with family members to improve digital wellness. Apply each challenge for a set period and observe its effect on your mood, focus, and daily life.

- **A Day Without Your Phone:**

Set aside one day each week or month to be truly offline in every sense of the word. Choose a day off when you are unlikely to need to respond to work-related messages, and inform friends and family in advance that you will be unavailable for twenty-four hours. At first, you may experience an unfamiliar sense of emptiness or anxiety; this reaction is completely normal. However, once the initial discomfort passes, many people report a strong feeling of freedom and achievement.

You may be surprised by how much time was previously lost to unnecessary phone checking. Plan enjoyable alternatives for the day, such as exercising, visiting relatives, reading an entire book, or returning to a forgotten hobby. Research suggests that even one day away from a phone can significantly reduce stress and improve overall mood. This was also evident in the stories of Tia and Anas, for whom a single day helped restore calm and focus. If a full day feels overwhelming, begin with half a day and gradually increase the duration. At its core, this challenge reinforces one message: you are in control of your digital habits, not captive to them. At the end of the day, reflect on your experience. Did you truly miss your phone, or did your mind finally have room to breathe?

- **A Weekend Without Social Media:**

Are you willing to spend two consecutive days without scrolling through Facebook, Instagram, or X? This challenge targets one of the most time-consuming and tempting digital habits. The following weekend, try deleting social media apps temporarily, or at least logging out from Thursday evening until Monday morning. This approach resembles a social media fast, often described as a weekend digital detox. During these two days, you may notice how many hours suddenly become available, hours once absorbed by endless scrolling.

Use this reclaimed time for more fulfilling activities. Spend extended time with family, engage in sports, go on outings, volunteer, or visit friends. One mother who tried this challenge with her family reported that they felt more connected after a weekend away from screens. You can also use the time to explore a new hobby or complete tasks you have been postponing. While the challenge may reveal how deeply social media is embedded in daily life, its most powerful outcome is realizing that life continues smoothly without it, and that news and updates can wait. After succeeding once, you may choose to repeat the challenge monthly or adopt regular digital breaks for mental rest.

- **A Device-Free Zone:**

Create a simple sign or rule that encourages family members to put devices aside during moments of real connection, such as meals or shared activities. Establishing a device-free space reinforces presence and interaction. Combined with a social media-free weekend, this practice allows families to rediscover the joy of shared experiences without screens competing for attention.

- **A Digital Mindfulness Hour:**

This simple daily challenge can have a powerful impact. Set aside sixty minutes each day that are completely free of electronic devices, and dedicate this time to your mental wellness. You might choose early morning hours to begin the day calmly, practicing mindfulness, prayer, or journaling while enjoying a warm drink without interruption. Alternatively, you may prefer an evening hour before sleep, using the time for relaxation, reading a physical book, or engaging in meaningful conversation.

The key is consistency. Choose a time you can commit to daily and allow yourself to sit with your thoughts without distraction. Some people describe this practice as a daily mental detox that restores balance amid constant digital stimulation. You can also use the hour for digital reflection, reviewing how you used technology during the day: what worked well, what did not, and how you can improve tomorrow. This pause strengthens self-awareness and helps restore control over habits. Specialists note that regular mindfulness and time in nature away from devices protect mental health and reduce stress. Treat this hour as sacred, and over time you may notice greater calm and positivity throughout the day.

- **A Week Without Notifications:**

If turning off your phone entirely feels too challenging, try a less intense yet highly effective option: spend a full week without non-essential notifications. At the beginning of the week, disable alerts for apps such as social media, games, and shopping platforms, keeping only truly important notifications, such as urgent work messages or close family contacts if necessary. Mute sounds and vibrations as much as possible so that your phone remains quiet unless something essential occurs.

You may still open apps intentionally at scheduled times, such as checking messages twice a day, but you will no longer be reacting to constant interruptions. Throughout the week, observe how your focus and emotional state change. Many people report a noticeable sense of calm and improved concentration once notification overload disappears. Using features such as Do Not Disturb mode during work and rest hours can further support this process. At the end of the week, reflect on the experience. Did you truly miss anything important? In most cases, the answer is no, while the gain in peace of mind is significant. Afterward, you may choose to extend the challenge or permanently keep most notifications disabled. This practice helps reclaim control of attention and prevents it from being fragmented across countless apps.

These challenges, along with many others you can easily invent, such as a daily family hour with no electronics or a week without YouTube videos, can be powerful tools for change. The key is to approach improving your digital habits with a spirit of experimentation and play, rather than punishment or deprivation. Treat these challenges as opportunities to explore a healthier lifestyle, not as restrictions. Reward yourself and your family after completing each challenge. Go out for a nice meal together or celebrate reaching the goal in a simple, joyful way. Over time, these short-term challenges can naturally turn into lasting habits without requiring much effort. As one technology expert wisely noted, "Even a gradual reduction in use, by just two hours a day, can give you back an entire month of your life every year."

The Impact of Digital Wellness on Mental Health, Relationships, and Productivity

Adopting practical digital wellness guidelines does more than provide temporary relief from phone-related distraction. It creates deep and lasting improvements across our psychological, social, and professional lives. The following overview highlights these effects, drawing on research findings and real-world experiences.

Mental Health and Emotional Balance

It is increasingly clear that reducing excessive screen time is closely linked to better mental health. When device use is limited and distractions such as constant notifications are removed, daily levels of anxiety and stress tend to decline. In one study, researchers found that abstaining from phone use for just one day led to a noticeable reduction in stress hormones among participants. This occurs because the brain finally gets a break from the continuous state of alertness created by nonstop communication and the endless flow of online information.

Remaining digitally present at all times, while consuming constant streams of news and social content, drains the mind and exhausts emotional energy through comparison, pressure, and exposure to negative material. When healthy digital boundaries are set, many people experience a sense of inner relief, as if a heavy weight has been lifted. Individuals who have tried digital detox practices often report improved mood and increased feelings of happiness after only a few days of reduced use, along with lower levels of depression and psychological strain. Sleep quality also improves, as reduced screen exposure, especially before bedtime, supports deeper and more restorative rest. Blue light and mental stimulation before sleep are known contributors to insomnia and disrupted sleep patterns. For this reason, digital wellness is inseparable from mental health. By organizing your relationship with technology, you gain calmer emotions, sharper focus, and a stronger ability to cope with daily pressures.

Social and Family Relationships

It is often said that technology has brought distant people closer, but pushed those nearest to us further away. Unbalanced technology use has a direct impact on the quality of human relationships. How often do family members sit in the same room while each person is absorbed in their own screen? These common scenes reduce warmth and create emotional distance, even when people are physically together. Applying digital wellness principles helps correct this imbalance and restore closeness.

When families set device-free times, conversation, laughter, and genuine connection return. In the experience of the mother who introduced device-free time, she reported that for the first time in a long while, she heard meaningful details from her children about their lives that had previously gone unspoken. Anas also observed during his phone-free day that stepping away from digital noise allowed him to reconnect with himself, which later helped him connect more deeply with others.

Marital relationships can similarly benefit when partners agree on phone-free times or spaces. Instead of scrolling separately in the evening, couples can talk, share activities, and strengthen understanding and emotional closeness. In contrast, studies show that excessive phone use, particularly for work outside official hours, increases tension at home and negatively affects family relationships. Establishing a daily disconnection rule after a certain time helps restore each person's right to personal and family life. Even friendships improve when face-to-face interaction replaces constant virtual communication. Meaningful memories are created through shared experiences, not through accumulating digital likes. Digital wellness strengthens relationships because it allows us to be fully present, to listen attentively, and to show care without interruption. As one specialist observed, "Having devices in plain sight during personal interactions reduces their quality." Simply keeping the phone out of view can make conversations deeper and emotions clearer.

A family enjoying time together without digital distraction, where communication improves and shared joy increases. Setting device-free time strengthens family bonds and creates lasting memories beyond the virtual world.

Productivity and Focus at Work and in Study

It is no secret that digital distraction has become one of the greatest enemies of productivity in modern life. Constant notifications and frequent task switching fragment the attention of both employees and students. Research shows that every interruption, whether from a notification or a quick glance at a screen, breaks concentration and may require up to twenty minutes to fully return to the original task. For this reason, implementing digital wellness practices, such as disabling non-essential notifications and scheduling specific times to check email, leads directly to improved focus, efficiency, and overall performance.

In France, lawmakers justified the "right to disconnect" by explaining that the pressure of continuous electronic communication after working hours led to employee exhaustion and reduced productivity. Globally, companies are increasingly realizing that allowing employees genuine breaks from constant communication makes them more creative, more focused, and more effective during official working hours.

For students, balanced technology use is also closely linked to stronger academic achievement. When a student sets the phone aside while studying, they can often absorb in one hour what might otherwise take two hours with the phone nearby, due to attention drifting between paragraphs and notifications. One study found that reducing time lost to aimless browsing allows individuals to complete core tasks more efficiently and in significantly less time.

In addition, setting dedicated work periods without digital interruptions improves the quality of output. For example, a report can be completed in two hours of deep focus instead of stretching into four fragmented hours interrupted by notifications, meetings, and messages. Creativity also benefits from digital wellness, because the mind needs quiet, uninterrupted periods where ideas can develop freely. For this reason, many writers, designers, and thinkers recommend going offline for a period during brainstorming or creative work.

From this perspective, digital wellness becomes an investment in professional and academic success. By reducing distractions, you gain time, clarity, and higher achievement. You will not only feel the difference yourself, but your manager or teacher will likely notice improvements in your focus and the quality of your work. At the same time, this balance reduces digital burnout, the deep exhaustion caused by constant screen exposure, helping protect your energy for sustained productivity and creativity over the long term.

Physical and Health Benefits

It is also important to highlight the physical health benefits of digital balance. When we consciously reduce excessive screen time, we decrease problems such as eye strain and dryness caused by prolonged staring, as well as neck and back pain resulting from poor posture while using phones and laptops. Movement during digital breaks stimulates blood circulation and helps prevent the physical stiffness and fatigue caused by sitting for long periods.

On another level, using devices less frequently reduces the tendency to snack mindlessly while scrolling or to stay up late in front of screens, both of which indirectly affect weight and overall health. Recent studies have also linked immunity and mental wellness to screen exposure. Excessive use can contribute to sleep problems and gradual weight gain, while moderation and balance support a healthier lifestyle overall.

In short, digital wellness is a gateway to comprehensive wellness. Improving your relationship with technology positively affects your mind, emotions, relationships, work, and physical health. It is a small stone that creates wide, positive ripples across the lake of your life. When you begin to feel these changes, you will realize that the effort required to adjust your digital habits was well worth it.

Recommendations for Creating a Personal Digital Wellness Plan

After reviewing the strategies and benefits, you may be wondering how to create a personal plan that truly suits your life. The following practical steps can help you build a customized digital wellness plan and put it into action, starting today.

1. Assess Your Current Situation and Raise Awareness

Begin by understanding your starting point. Track your digital habits for two or three days and take notes. How many hours do you spend on your phone each day? Which apps or websites consume most of your time? At what times do you use devices the most, in the morning, before bed, or during meals? Many people are surprised once they see the numbers clearly.

Use your phone's built-in Screen Time or Digital wellness reports, which provide detailed weekly summaries of usage. You can also ask family members for their observations. They may notice patterns you overlook, such as frequent evening scrolling or missed moments of connection. The purpose of this step is to increase self-awareness, understand how technology affects your life, and identify the areas that most need attention.

2.   Set Clear, Measurable Goals

Next, define what you want your plan to achieve. Make your goals specific and measurable. For example, reduce daily screen time from six hours to three hours within one month, avoid phone use during dinner every day, sleep before 11 PM without using your phone, or spend two device-free hours each week with your family.

If this is a family effort, involve everyone in setting shared goals, such as agreeing on a household rule like "no phones at the dinner table." Ensure your goals are realistic and gradual. Avoid vague intentions like "I will stop using the internet," which are neither practical nor sustainable. Instead, understand your current habits and aim for steady improvement.

You can also divide goals into phases. For example, a short-term goal for the first week might be turning off notifications at night, while a longer-term goal for three months could be making Friday a social media–free day. Clear goals give your plan direction and allow you to track progress and adjust when needed.

3.   Build New Routines and Boundaries

Transform your goals into daily and weekly routines by creating clear rules and boundaries. If your aim is to reduce phone use, your plan might include no phone use after 9 PM, limiting browsing sessions to 20 minutes with a timer, or keeping the phone on silent during study or work hours.

If strengthening family time is your goal, establish device-free zones or time windows. For example, decide that phones are not used in the living room between 7 and 9 PM, when the family gathers. Posting a friendly reminder note in that space can help everyone remember and cooperate.

Experts emphasize the importance of separating work from leisure by designating times and spaces where devices are not allowed. You might include rules such as "no work email after 6 PM" or "phones stay outside the bedroom." If sleep improvement is a priority, design a consistent evening routine that requires turning off electronics well before bedtime.

These boundaries form the core of your plan. Write them down clearly and keep them visible as daily reminders.

4.   Create Healthy Alternatives and Parallel Activities

Reducing screen time without knowing how to use the freed time often leads to frustration. Therefore, prepare positive alternatives in advance. Think about what you enjoy or would like to explore instead of using your phone. This could include reading, learning a new skill, exercising, drawing, or spending quality time with family.

Schedule these activities so that when a device-free moment arrives, you already have a plan. For example, take a 30-minute walk on Tuesday and Thursday evenings, play sports with your children on Saturday morning, or host a family movie night on Friday without phone checking during the film. If your goal is to reduce online social interaction, plan regular face-to-face meetings with friends or relatives.

If you feel attached to a particular app, try replacing it with an activity that meets a similar need. For instance, replace one hour of social media scrolling with learning a new recipe or watching a meaningful documentary. Entertainment is not the problem; unplanned excess is. Make technology the last option rather than the first response to boredom, and be creative in building motivating alternatives.

One mother even introduced a "time bank" system for her children. She tracked the hours they spent on beneficial activities such as reading, sports, and homework, and allowed them equivalent screen time later. This approach increased motivation while keeping total usage balanced.

5.   Involve Family or Friends and Build Mutual Support

Digital habits rarely change in isolation. We influence those around us, and they influence us in return. Whenever possible, turn your digital wellness plan into a shared project.

Explain your goals to your family and ask for their support. You might agree on shared challenges, such as competing to reduce screen time the most. If you have children, involve them in setting rules. Young people are far more likely to follow guidelines they helped design.

Ask questions like, "What do you think is a reasonable amount of screen time each day?" or "What activities would you like us to do together instead?" This approach makes children partners in change rather than passive recipients of rules.

You can also invite a close friend to act as an accountability partner. For example, agree to turn off phones at 10 PM and send each other friendly reminders. Research shows that group-based efforts make behavior change easier and more sustainable. In digital detox efforts, social support reduces isolation and strengthens commitment.

Do not hesitate to share your intention to improve your digital habits. You may find that your example encourages others to reflect on their own use.

6.   Using Supportive Technologies (Yes, Technology Can Help Regulate Technology)

Although it may seem contradictory, technology itself can support digital wellness when used wisely. Most smartphones offer advanced tools for managing screen time. Features such as Focus Mode on modern phones allow you to create specific settings for work, sleep, or relaxation, permitting only essential apps while silencing the rest.

Used intentionally, these tools can help you stay aligned with your goals rather than distracted by constant digital noise.

Set up Focus modes to align with your plan. For example, Sleep mode can block everything except alarms and emergency calls, while Work mode can silence entertainment and social media apps. These small adjustments significantly reduce unnecessary interruptions.

There are also apps designed to support mindfulness and relaxation. You can use them intentionally, then switch your phone to airplane mode to prevent disruptions. Some people even choose lock screen wallpapers that gently question their habits, with reminders such as: "Do you really need to open your phone right now?" If your work requires long hours on a computer, time-tracking tools can also be helpful. These tools show where your time goes and alert you when too much time is spent on unproductive activities.

Another effective step is reorganizing your phone. Place work, learning, or productivity apps on your home screen, and move social media platforms and games to a distant screen or a hidden folder to reduce the habit of automatic tapping. You may also experiment with switching your screen to grayscale for a period. Without bright colors and visual stimulation, apps become noticeably less tempting. In short, make technology your ally by adjusting settings and using tools that support balance rather than undermine it.

7.   Gradual Change, Patience, and Adjusting

As you begin implementing your plan, remember that you are changing habits built over many years. Patience and realism are essential. You may not succeed perfectly in the first week, and you may break some rules at the beginning, and that is completely normal. What truly matters is consistency, not perfection.

Adopt a gradual approach. If you currently spend eight hours a day on screens, do not expect to drop to two hours overnight. Start by reducing two hours in the first week, then one more hour in the following week, and continue steadily. As Wall Street Journal writer Julie Jargon notes, cutting just two hours of screen time per day can reclaim almost an entire month of time over the course of a year.

Focus on small daily changes rather than large decisions all at once. For instance, begin by banning phones at the dinner table today, then introduce a no-phone-in-the-bedroom rule tomorrow. If a particular strategy feels unrealistic or overly difficult, adjust it without viewing the change as a failure. Your plan is not rigid; it is designed to evolve as you learn more about your habits and needs.

Tracking progress weekly is especially helpful. At the end of each week, reflect on what worked and what was challenging, either on your own or with your family if they are involved. Ask yourself whether you noticed benefits such as better sleep, longer conversations, or improved focus. Use these observations to guide your goals for the following week. This regular review process is what transforms a digital wellness plan into a sustainable lifestyle rather than a short-lived burst of motivation.

8. Continuous Rewards and Motivation

Finally, remember to reward yourself along the way. When you reach a goal or follow your plan consistently for a period of time, acknowledge your effort and celebrate your progress. Treat yourself to something enjoyable, preferably non-digital, or plan a relaxing outing. Every positive step brings you closer to a more balanced and fulfilling lifestyle.

Encourage your family members as well. If your child respects the agreed screen time limits throughout the week, offer praise or a simple reward, such as choosing a movie to watch together. Positive reinforcement sends a powerful message to the brain that this journey is both meaningful and enjoyable, strengthening motivation to continue.

Keep reminding yourself why you started. Whether your goal is improving health, deepening family connections, or increasing productivity, reconnecting with your purpose will help you stay committed. You might place a note on your desk listing the benefits of digital wellness or display an inspiring quote such as: "When you take control of your digital tools, you take control of your life."

By following these steps, you can design a personal digital wellness plan that suits your needs and circumstances. The key lies in balancing flexibility and commitment, flexibility to adapt when needed, and commitment to your intentions and goals. The process may feel challenging at first, but every meaningful change begins with small, consistent actions. Your plan is your roadmap, and you are the one leading the journey.

Toward Personal and Family Success Through Digital Wellness

As this chapter concludes, it is important to reaffirm that digital wellness is not a luxury or optional comfort. It is a modern necessity that enables us to navigate life with clarity, confidence, and health. Every practice discussed, from disabling unnecessary notifications to taking intentional digital breaks and engaging in motivating challenges, acts as a building block in a protective barrier against the harmful effects of uncontrolled technology use. Behind each block lies a broader goal: regaining control over time, attention, and the overall quality of life.

When digital balance is achieved, its impact extends into every area. Mental health improves, emotions stabilize, and the ability to enjoy the present moment grows stronger. Relationships become richer, conversations deepen, laughter returns, and the voices of loved ones become louder than the sound of notifications. Productivity increases as hours once lost to aimless scrolling are transformed into meaningful achievements and personal growth.

At the family level, adopting a culture of digital wellness helps children develop healthier entertainment habits and stronger self-directed learning skills. Family bonds strengthen, and the example set by parents shapes how children understand and use technology for years to come.

This approach does not call for abandoning technology or withdrawing from the modern world. On the contrary, digital wellness means engaging with technology thoughtfully while preserving its benefits. It is the ability to choose intentionally, organize wisely, and ensure that devices remain tools serving personal goals rather than forces controlling daily life. Those who use technology with awareness to support learning, work, and communication will experience real progress and success. Those who allow unchecked digital consumption may one day realize that valuable years have quietly slipped away.

Let digital wellness become a way of life. It may feel difficult at first, as all meaningful change requires effort, but you are not alone in this journey. Countless individuals around the world have recognized the need for balance and have succeeded, and some of their stories have been shared here. You can be one of them. Begin with small, steady steps, and the difference will soon become clear.

Finally, consider the larger picture. A digitally aware individual contributes to a healthier family and becomes a more creative, focused employee or student. Together, such individuals form a balanced society capable of using technology as a tool for growth rather than harm. Your personal and family success depends on creating this balance, and once achieved in your digital life, it will naturally extend into your real world.

Let digital wellness be your daily companion and constant reference point. Use this chapter as your guide and roadmap toward a responsible and fulfilling digital life. You are now holding the reins of your devices, moving forward with confidence and intention, and this balance is undoubtedly one of the keys to success in the twenty-first century. Wishing you every success on your journey toward sustainable digital wellness.

Chapter Nine:
"Inspiring Real-Life Case Studies"

In this chapter of the book "Digital Wellness," we present inspiring real-life examples of individuals, families, and organizations that have successfully improved their relationship with technology. We explore how young people, children, parents, and employees managed to change their digital behavior for the better, and how families introduced digital wellness rules that led to clear, positive results.

We also highlight successful initiatives in schools and universities across the Gulf and around the world, as well as the efforts of institutions and companies that have embedded digital wellness into their organizational culture.

These stories are presented as case studies that clearly illustrate the "before" and "after" of each experience: What challenge was faced? What action was taken to create change? And what positive outcomes followed? Wherever possible, we include the voices of the people behind these experiences, either through direct quotes or through insights shared by specialists who observed their journeys.

The purpose of this chapter is to inspire the reader, showcase the practical and human side of digital wellness, and emphasize that positive change is achievable everywhere, from the UAE, Saudi Arabia, and Egypt to Canada, Sweden, and Japan. These real-world models demonstrate how spreading a culture of digital wellness can truly transform daily life.

Stories of Individuals Who Changed Their Digital Behavior

Real change begins with the individual. In this section, we share stories of young people, children, parents, and employees who recognized that excessive technology use was negatively affecting their lives and decided to take courageous steps to reshape their digital habits. Each story explores life before the change, marked by digital dependency and its consequences, the actions taken to regain control, and the positive outcomes achieved in mental health, relationships, and personal success.

Violet's Story: A Teenager Overcoming Phone Addiction

From the United Kingdom comes the story of Violet, a 14-year-old girl who once spent most of her day on her smartphone. Before making any changes, Violet spent up to ten hours daily scrolling through apps and social media, which left her feeling stressed and increasingly disconnected from her family. She began to grasp the seriousness of the problem when she realized how much of her time was disappearing into her phone, leaving little space for studying or hobbies. Violet recalls, "I wanted to reduce my screen time, but I kept asking myself, how can I do that when my phone feels like everything?" This question became the starting point of her challenge.

Action taken: Violet joined a school challenge that lasted three weeks and focused on reducing daily screen time. She set firm rules for herself, limiting Snapchat to one hour per day and TikTok to two hours, and avoiding usage beyond those boundaries. She also filled her free time with enjoyable offline activities, such as baking and reading.

"I managed to read four books this month," she proudly says, highlighting how she replaced hours of scrolling with a habit she had long neglected.

Outcome: Within just a few weeks, Violet reduced her daily phone use from ten hours to less than four. This dramatic change had a visible impact on her mental wellness and her relationship with her family. She felt calmer, happier, and more emotionally balanced, and she became noticeably more engaged at home. Her father, James, reflects on the transformation, saying, "I used to believe it was normal, even necessary, for children to spend long hours on their phones to keep up with the times. I didn't realize how much it was affecting my daughter's wellness until I saw the change."

Violet's parents and grandparents noticed that she became more involved in family activities and less isolated. She began helping with cooking and joining weekend outings instead of remaining absorbed in her phone. She also developed new interests, including collecting classical music records, alongside her renewed love of reading. Violet explains that her relationship with her phone shifted from dependency to conscious, limited use.

Her experience inspired other teenagers at her school, and several classmates expressed a desire to follow her example after seeing its positive effects. Violet's story clearly illustrates the power of change, showing a transformation from a phone-dependent teenager struggling emotionally and socially into a more balanced young person who is happier and more connected to real life.

The Story of a Saudi Young Professional Finding Balance Between Work and Digital Life

Digital overuse is not limited to teenagers. Young professionals can also find themselves immersed in screens for most of the day. From Saudi Arabia, we present the case of a young employee in his late twenties, whom we will call Yusuf, working in the digital marketing sector.

Before the change: Yusuf lived in a constant state of connection. His workday revolved around a computer screen, his breaks were spent checking his phone, and even after office hours he remained available for messages and emails at all times. "I felt as if there was a switch inside me that was always on and never turned off," Yusuf explains. This lifestyle led to exhaustion, reduced concentration, and a growing sense that life beyond the screen was slipping away. His main challenge was breaking the cycle of nonstop connectivity and restoring balance between work and personal life.

Action taken: Yusuf began with small but bold steps toward digital balance. One day, he went for a walk without carrying his phone, something he had not done in years. Anxiety immediately followed. "After an hour without my phone, I felt uneasy. What if someone tried to reach me? What if something happened and I needed help?" he recalls.

As he repeated the experience daily, the anxiety faded and was replaced with a sense of freedom. "When I focused on walking and observing the road, I felt like I had regained something I had lost," he says.

Gradually, Yusuf added clearer boundaries. He stopped taking his phone into the bedroom to protect his sleep and banned phones from the dining table during family meals. He also set a firm evening cut-off time, switching his phone off completely at 9 PM. To reinforce this boundary, he placed a message on his WhatsApp profile stating, "I turn off my phone after 9 PM." This public commitment helped him stay disciplined. Once the time arrived, he turned his phone off without hesitation and enjoyed uninterrupted personal time.

Outcome: Within weeks, Yusuf noticed a remarkable improvement in his mood and daily life. During his phone-free walks, he rediscovered simple pleasures he had overlooked, such as watching sunsets and noticing seasonal changes. "I started seeing leaves fall and realizing that phones are not an end in themselves, but simply a tool," he reflects.

Evenings became calmer and more meaningful. He spent quality time with his wife, read books, and engaged in hobbies instead of staring at a screen. His sleep quality improved significantly, and he began going to bed with peace of mind, no longer chased by notifications.

Socially, his relationships deepened. "When we meet, conversations feel more real when I ask people directly about their lives instead of following their updates online," Yusuf explains. Professionally, his productivity did not decline. In fact, he arrived at work more focused and energized.

Yusuf's story demonstrates a clear before-and-after transformation: a young professional drained by constant connectivity becoming a calmer, happier, and more balanced individual through simple daily boundaries. His experience also encouraged colleagues to adopt the idea of disconnecting after work, reminding them that rest is a right, not a weakness.

A Mother and Her Two Children: A Family Experience in Breaking Free from Screens

In Australia, a mother undertook an inspiring experiment with her two children, aged six and eight, by committing to a full week of digital detox from electronic devices. Before this decision, the family had not enforced strict screen time limits, and the children were accustomed to spending after-school hours watching television or using tablets to play games and watch videos. Over time, the mother noticed a troubling pattern: the children began expecting screens even during social moments. When relatives or friends visited, they preferred retreating with their devices rather than interacting.

At that point, the parents realized a decisive intervention was needed to break the children's screen dependency and guide them back toward age-appropriate, natural activities.

The challenge: The mother and her husband decided to implement a complete digital cut-off for one full week for the entire family. They knew the first few days would be the most difficult and fully expected resistance, boredom, or complaints from the children. To prepare, they planned enjoyable alternatives in advance, including outdoor play, park visits, board games, and bedtime story reading instead of cartoons.

Action taken: At the beginning of the week, all tablets and phones were collected and placed out of the children's reach, and the home internet was temporarily switched off at certain times to minimize temptation. The family agreed on one clear rule: "No screens at all this week," and committed to following it together. Importantly, the parents also avoided using their own phones in front of the children, setting a consistent and credible example.

During the first days, the children experienced moments of boredom and repeatedly asked for their devices. However, the mother kept them actively engaged. She took them to the park to play ball, involved them in baking a cake in the kitchen, and set aside time for drawing and coloring together. Gradually, the children adapted to the new routine. The mother admits she was surprised by how quickly they adjusted. After the second day, the children stopped asking as often, began inventing games together, and filled their time with imaginative free play.

Outcome: The results were striking. The mother observed noticeable improvements in her children's behavior throughout the week. "The children were not upset at all by the absence of technology," she explains. "On the contrary, they were calmer and in a better mood." The usual conflicts before school or bedtime disappeared, as there were no sudden device shutdowns triggering frustration in the morning or at night.

The daily routine became simpler and more natural. As soon as the children returned from school, they went straight outside to play in the garden and with the family dog instead of heading toward a screen. Over the weekend, the family enjoyed outdoor activities such as bike riding and playing ball, and the home atmosphere was filled with energy and laughter rather than the artificial silence that once dominated when everyone retreated into separate screens.

The mother admits that she personally benefited as much as her children. Although the detox began as a way to support them, she realized that she, too, had been spending excessive time on her phone. By reducing her own phone and television use, she felt less stressed and more productive in both household responsibilities and professional work. She began reading before bed instead of scrolling and incorporated light exercise into her evenings, which positively affected her sleep and overall wellness.

At the end of the week, the family gathered to reflect on the experience and agreed to adopt long-term rules, such as limiting daily device use and never allowing screens during meals or when guests are present. The mother reflects, "I felt like, as a family, we gained back a week of our lives. We talked, laughed, and moved together more than we had in a long time."

Their home transformed from a place that appeared peaceful only because everyone was absorbed in a screen into a space filled with movement, interaction, and genuine connection. This short experiment motivated the family to keep technology in its proper place and prevent it from dominating daily life.

Lessons Learned Summary (Individuals)

These stories demonstrate that individual change is both achievable and deeply rewarding. The teenager who reduced her digital dependency gained balance and confidence. The young employee who set firm boundaries around technology felt he had regained control of his time and strengthened his relationships. The family that committed to a week without screens rediscovered connection, calm, and shared joy. The common thread across these experiences is willpower combined with small, realistic steps. Simply setting the phone aside for an hour a day, switching it off after a fixed time, or committing to a device-free weekend can create meaningful change. These real-life examples can inspire anyone who feels overwhelmed by technology to take a first step, whether small or bold, toward healthier digital wellness. Positive results often appear quickly, whether in improved mental health, better sleep quality, or stronger social relationships.

Families Adopting Digital Wellness Rules

After exploring individual initiatives, we now turn to the family level. In many cases, the family represents the first line of defense in achieving digital balance, especially when children are involved. In this section, we present case studies of families from diverse cultural backgrounds, ranging from the Gulf and Egypt to North America and Europe, who introduced clear rules and boundaries for digital use at home. These changes produced noticeable improvements in family bonding and personal wellness. We examine the challenges these families faced, the strategies they adopted, and how daily life shifted before and after implementation.

Family Rules: The Challenge, the Solution, and the Outcome

The Challenge: In an era where devices are present in every corner of the home, many families struggle with screens dominating communication and interaction. Common scenes include children sitting silently at the dinner table absorbed in phones or tablets, and teenagers isolating themselves in their rooms instead of joining family activities. In some households, parents openly complain about the absence of real conversation. Studies indicate that 33% of parents in Saudi Arabia rely on technology daily to occupy their children rather than engaging with them directly, and that excessive screen use has reduced family interaction time for 42% of people. These figures highlight the scale of the issue: technology has gradually replaced warm family moments.

As a result, many families recognized that change was necessary and that new household rules were required to limit device domination and restore balance.

Action Taken:

- *Family digital wellness rules:* Families that successfully improved their digital wellness adopted rules that were simple yet firm. Key examples include:

- *Screen-free zones and screen-free times:* Many families designated specific areas or moments where devices were not allowed, such as the dining area during meals or bedrooms before sleep. One mother introduced a strict rule of no phones at the dinner table, allowing conversation to take priority during meals.

- *Night-time device curfews:* Some families agreed to switch off all devices at a fixed evening time, such as 10:00 p.m. This helped children return to healthier sleep routines and gave parents uninterrupted time to rest or connect.

- *Shared alternative activities:* Families deliberately introduced enjoyable group activities to replace screen time, including weekly board-game nights, evening walks, or shared reading sessions. One family in Canada established a Friday evening with no electronics, during which they watched a movie together, played chess, and strengthened their emotional bonds.

- *Parents leading by example:* Across all successful cases, one rule stood out as essential: parents must model balanced digital behavior themselves. Asking children to reduce screen use while adults remain glued to their phones undermines any rule. For this reason, parents committed to limiting their own device use, especially during shared family time.

Results: The outcomes of these family rules were overwhelmingly positive. In the UAE, families who participated in a 21-day digital wellness challenge reported significant improvements. Parents noted up to a 37% improvement in sleep quality after reducing screen exposure before bedtime. Families spent more time together without digital interruptions, and many described having deeper conversations than they had experienced in years. Notably, 100% of participating school-aged children said they felt proud of successfully managing their own screen time by the end of the challenge. After a family workshop on digital use, one mother shared a powerful reflection: "We have not spoken this openly in years. I feel like I am seeing my son again." This statement captures how honest conversations about technology can dissolve emotional distance and rebuild meaningful family connection.

Other visible outcomes included clear improvements in children's behavior and academic performance after the new rules were introduced. For example, once a night-time device switch-off policy was adopted, children began waking up with more energy and noticeably less irritability. This change had a direct and positive effect on their concentration and performance at school. Parents also observed that free time previously lost to screens gradually shifted toward meaningful hobbies. One child developed a passion for drawing, another took up cycling, while a third began cooking with his mother every evening. These simple yet powerful lifestyle changes emerged merely from resetting digital priorities within the home.

It is important to note that these family successes did not come without challenges. The first few device-free evenings were often difficult, marked by boredom, resistance, or mild tension. However, families that remained consistent and encouraged patience were able to move beyond this initial adjustment phase. Over time, everyone adapted to the new routine and began to appreciate its benefits. Many parents expressed surprise at how quickly children adjusted when enjoyable alternatives and strong family support were provided. In short, the keys to success were consistency and shared commitment: applying the rules firmly without backtracking, and ensuring that every family member participated through encouragement, accountability, and follow-through.

A Model from an Arab Context: The "Amani Dot Com" Campaign in Egypt

Efforts to promote digital wellness are not limited to individual or family initiatives. At a broader level, national campaigns have also played an important role in supporting families. In Egypt, for instance, the government, in cooperation with UNICEF, launched a digital awareness campaign under the slogan "Amani Dot Com," aimed at protecting children and enhancing their wellness in the digital world. In its first phase, the campaign focused on raising awareness among children, parents, and caregivers about the foundations of safe and balanced internet use. It also provided practical guidance on protecting children from cyberbullying and other online risks. Workshops and digital training camps were organized to help families learn how to set household rules for online use and how to report abusive or harmful content that children might encounter.

This national Egyptian initiative offered a supportive and accessible model for families seeking to adopt digital wellness practices. By presenting easy-to-understand content and practical advice, the campaign empowered parents to create a safe and healthy digital environment at home. It also achieved strong engagement on social media through the hashtag #AmaniDotCom, where many parents shared their experiences of applying the campaign's recommendations and setting clear limits on their children's device use.

Through these diverse family models, it becomes clear that the family's role is pivotal in establishing a culture of digital wellness. Parents are the ones who set the rules, draw boundaries, and act as daily role models. The earlier a family teaches a child how to balance technology use, the more likely that child will grow up able to enjoy its benefits without falling into its negative consequences. As the saying goes, "digital upbringing starts from the cradle," meaning that instilling healthy habits early makes the process far easier later in life. At the same time, it is never too late for families struggling with digital overload to begin restoring balance. These experiences clearly demonstrate that simple steps, combined with collective commitment, can lead to tangible and lasting positive results.

School and University Initiatives to Promote Digital Wellness (in the Gulf and Worldwide)

The responsibility for spreading digital awareness does not rest solely with individuals and families. Schools and universities also play a fundamental role in shaping the habits of the next generation. Across the Gulf and around the world, innovative educational initiatives have emerged to raise students' awareness of digital wellness and encourage healthier technology use among young people.

In this section, we review inspiring examples from educational institutions, ranging from a school-based program in the UAE that brings students and parents together to achieve digital balance, to an innovative university course in the United States that addresses the topic through interactive learning. We also examine school policies in Europe that have proven effective in reducing students' dependence on smartphones. Together, these examples illustrate how educational institutions can produce students who are more aware, responsible, and intentional in the digital world, while serving as strong partners to families in supporting children's digital wellness.

UAE: The Digital Bridge Project and the 21-Day Challenge in Dubai Schools

The UAE witnessed a pioneering experience considered one of the first of its kind to target both schools and families simultaneously. This initiative took the form of a community-based research project known as the "Digital Bridge," launched by EdRuption in collaboration with several schools in Dubai. The project aimed to reset digital use habits among students and parents through a structured 21-day challenge.

Before the Project: Observations revealed that many students were spending excessive amounts of time on screens in their daily lives. There was also a noticeable gap between parents' understanding of the role of technology in their children's education and the actual ways in which children were using it. In addition, school policies regarding electronic devices varied widely, and in some cases lacked meaningful involvement from students and parents. This situation formed the central challenge the "Digital Bridge" project sought to address: how to ensure that digital transformation in education remains balanced and healthy through genuine collaboration between school and home.

The Action Taken: Five diverse international schools in Dubai participated in the project, representing more than 8,000 students from over 100 nationalities. This cultural diversity gave the initiative strong momentum and relevance. The program began with a series of interactive workshops attended by parents and children together.

This joint participation was a key and uncommon feature, as families sat side by side to draft a shared digital charter.

Rather than relying on traditional lectures, the workshops used open dialogue and hands-on activities that encouraged participants to acknowledge challenges and propose realistic solutions. Each family created a digital pledge that included agreed-upon rules, such as specific times for device use, device-free zones like bedrooms, and a weekly family activity without screens. Following this, the 21-day challenge began, during which families implemented their pledges at home. Schools supported the initiative through regular encouragement, reminders, and reinforcement activities during the school day.

In addition, dedicated sessions were held for teachers and school leadership teams to help them become positive digital role models. Educators were encouraged to reflect on their own habits, such as phone use in hallways or responding to emails late at night, and to adjust these behaviors. This step recognized that adult actions strongly influence students' attitudes and awareness.

Results: The outcomes of the "Digital Bridge" project were both impressive and well documented. Post-challenge surveys and interviews revealed several key findings. Many parents reported improved sleep quality, with some experiencing up to a 37% improvement after adopting nightly screen-free routines. Parents also noted spending more uninterrupted time together as families, often describing the deepest conversations they had shared in years.

Among students, 100% reported feeling proud of their ability to manage screen time independently by the end of the challenge. This sense of achievement strengthened their self-confidence and motivation to continue healthy habits. Many students realized they could enjoy life without constant phone use and began exploring new hobbies.

The project also highlighted a generational gap in perspectives. While some school administrators favored total phone bans, students expressed a preference for balanced use rather than complete restriction. They wanted to be involved in creating the rules rather than having them imposed. Another notable finding was that 81% of participating parents were between 41 and 50 years old, bringing different generations together for meaningful dialogue and mutual understanding.

Beyond statistics, the project produced powerful personal stories. One father shared, "I have never felt this close to my son before. We spoke honestly about his digital anxieties and expectations, and now I understand that boundaries are a shared responsibility." A middle school student reflected, "At first I thought my parents were overreacting, but after the challenge I felt the difference. I slept better, focused more, and connected with my friends face to face."

These testimonies reflect a deep shift in attitudes, transforming digital wellness from a rule imposed by authority into a shared belief. The project's success demonstrated that sustained cooperation between schools and families is essential for long-term impact. Many families continued applying what they learned, and several schools adjusted their technology policies by involving students in creating digital use charters each term.

As the initiative expanded to other schools across the UAE, it became a model that other Gulf countries can learn from. Challenge-based, time-bound programs like this show how entire school communities can work together to achieve lasting improvements in digital wellness.

United States: A University Curriculum to Educate Adolescents About Digital Wellness

At the University of Michigan in the United States, an innovative educational initiative emerged to address digital health among adolescents by building a bridge between the university and middle schools. A team of professors and university students launched the "Digital Wellbeing for Young Adolescents" program, which brings together sixth-grade students, typically aged 11 to 12, with university students in an interactive learning environment. This unique initiative was designed as a peer-to-peer bridge between two generations of learners. Its goal is to raise younger students' awareness of the risks of digital addiction in a way that feels relatable and authentic, while also training university students to communicate effectively and responsibly about this topic.

The Program Idea and Its Challenges: University faculty observed that many digital awareness programs in schools rely heavily on a top-down, lecture-style approach, where adults instruct children through rigid lists of "do's and don'ts." This method has shown limited long-term impact. While students may listen to digital safety advice during formal lessons, it often fails to lead to genuine behavioral change or a deeper understanding of how technology affects their wellness. The core challenge, therefore, was finding a way to help adolescents engage meaningfully with the subject and understand it through both scientific insight and real-life experience.

The Action Taken: The solution was to actively involve university students from fields such as education, information technology, and social work in delivering workshops to younger students. The smaller age gap made conversations feel more natural, and the language used was far more relatable. To support this approach, the university created a course titled "Digital Wellbeing, Peer Communication." Undergraduate and graduate students enrolled in the course studied both theory and practice. The theoretical component explored current research on how digital addiction affects the brain and mental health, while the practical component involved visiting local middle schools to conduct interactive sessions with sixth-grade students.

These sessions included open discussions in which younger students shared their experiences with social media and electronic games, as well as simulation exercises demonstrating how digital platforms are designed to keep users engaged. For example, students learned about intermittent notifications and their psychological impact. The program also featured a group-based practical activity in which younger students tracked and analyzed their screen time. Over the course of a week, they recorded how many hours they spent on different apps and later discussed the findings with peers and university mentors.

Expert Involvement: The program was overseen by Professor Liz Kolb, a specialist in educational technology. She emphasized that the global health crisis brought on by COVID-19 intensified mental health challenges among young people due to the sudden and prolonged shift toward digital platforms. Kolb explained that as screens increasingly dominate learning, entertainment, and communication, there is an urgent need to teach digital wellness not through abstract warnings, but through scientific understanding. For this reason, the curriculum was designed to be rich in evidence-based content, including explanations of how late-night device use disrupts sleep and how constant comparison on platforms like Instagram can undermine self-confidence. Importantly, this information was presented in a simple, engaging manner that encouraged active student participation.

Results and Impact: The program was warmly received by both students and parents. Younger students responded enthusiastically to learning from mentors closer to their own age rather than traditional authority figures. One sixth-grade student, Cera, admitted that she enjoys watching short videos and reels for hours but struggles to stop. After participating in the program, she began to understand the design strategies that keep users engaged. She shared, "When I learned that designers intentionally try to make us addicted to these clips, I felt more motivated to control myself." Many students also began exchanging practical tips, such as using apps that alert them when they reach their daily screen-time limits.

For university students, the program offered valuable hands-on experience. Beyond earning academic practicum credits, they developed the ability to explain complex digital wellness concepts in language that adolescents could understand. One master's student reflected, "This program taught me that digital education requires psychological and cultural understanding of a child's generation. I will never forget the honest conversations we had about what students truly feel when using these apps."

This initiative highlights the power of innovation in digital awareness education. Instead of relying on traditional instruction, it adopted a peer-based, participatory learning model grounded in scientific research. An indirect but important outcome was increased parental engagement. At the program's conclusion, schools held a seminar inviting parents to view what their children had learned. Many parents were surprised by their children's honesty when they presented charts showing hours spent on platforms like YouTube or TikTok. Several parents admitted they would reconsider allowing unlimited smartphone access. In this way, the program influenced not only students but also sparked deeper conversations within families about digital habits.

Europe: Successful School Policies – A Case from Sweden

In Europe, several countries have taken bold steps at the policy level to reduce students' dependence on digital devices. Sweden offers a notable example. According to an official report released in 2024, the Swedish government described excessive screen time among young people as a "serious public health crisis" affecting both mental and physical wellness. Authorities noted increases in reduced concentration, obesity, and delays in the development of certain core skills, linking these trends to prolonged screen exposure. In response, the government advanced legislation requiring schools to ban smartphone use entirely during the school day up to a certain age. As a result, many lower and upper secondary schools introduced strict policies in which teachers collect students' phones at the beginning of the day and return them only at dismissal.

The Challenge and the Goal: The motivation behind this firm decision stemmed from concerning observations in schools. A principal in the city of Malmö described witnessing addiction-like behaviors, where students instinctively reached for phones even after they had been taken away. This unconscious action raised serious concerns about the depth of students' dependence. Teachers also reported declining classroom engagement and reduced physical activity, as students remained glued to their phones during breaks. The primary goal was therefore to restore focus in the classroom and protect students' physical health by encouraging movement, play, and face-to-face interaction.

The Action Taken: Swedish schools implemented collective phone collection systems using various methods. Some schools distributed fabric pouches in which students placed their phones each morning, retrieving them only at the end of the day. Others used lockers or secured storage boxes. To ease resistance, a few schools allowed limited phone access during lunch before collecting devices again.

At the same time, the Ministry of Health launched a public awareness campaign and issued clear screen-time recommendations for parents: no screen exposure for children under two, no more than one hour per day for ages two to five, a maximum of two hours for children aged six to twelve, and no more than three hours for teenagers. These guidelines aimed to ensure that homes reinforced the same principles applied at school.

Results: Testimonies from schools that implemented these policies over several years indicate clear positive outcomes. One deputy headteacher explained that within weeks of the ban, school playgrounds were filled with football matches, conversations, and laughter rather than silent groups staring at screens. Teachers also observed improved concentration during lessons, as constant digital interruptions were eliminated.

Perhaps the most unexpected response came from students themselves. Although initial resistance was common, many primary school students later admitted they felt freer under the new system.

In a report by DW, journalists interviewed 10- and 11-year-old students in Stockholm who openly acknowledged their struggles with overuse. One student said, "If the phone is close to us, we keep reaching for it. So it's better to keep it away so we can focus." Another added, "If my phone were with me in class, I wouldn't pay attention because the phone is more exciting."

Students also described coordinated efforts at home, with parents activating automatic device locks after two or three hours of use. This alignment between family rules and school policy is a key reason the Swedish model has proven effective. Schools limit phone use during the day, while parents manage screen time afterward.

Overall, Sweden's experience suggests that firm, health-driven policies can help restore balance to students' digital habits and protect them from serious risks. Critics argue that total bans may hinder the development of self-discipline, but supporters point to tangible outcomes, including reduced cyberbullying, increased physical activity, and stronger social interaction. While the policy continues to be reviewed and refined, many countries are now watching closely. UNESCO's 2023 report also recommended that schools consider restricting phone use to support academic focus. In this context, the Swedish approach stands as a strong case study that may inspire Gulf countries and others to develop similar measures tailored to their own educational and cultural settings.

A Culture of Digital Wellness in Institutions and Companies

The concept of digital wellness is not confined to personal life or educational settings. It also extends deeply into workplaces and organizations. As digital transformation accelerates across the business world, employees have encountered new challenges related to work–life balance, including emails that follow them after office hours, virtual meetings that stretch into the evening, and constant connectivity that can lead to digital burnout. In this section, we highlight examples of institutions and companies that recognized the importance of employees' digital wellness and chose to embed it within their workplace culture, whether through formal policies or voluntary initiatives. These cases show how such steps led to higher employee satisfaction and improved productivity, confirming that digital comfort is not a luxury, but a critical success factor for organizations themselves.

Shorter Workweek: The Microsoft Japan Experiment

One of the most widely cited real-world workplace experiments came from Microsoft Japan, where the company tested a four-day workweek by granting employees one additional paid day off each week during August 2019. The initiative was launched under the name "Work-Life Choice Challenge." The problem it sought to address was clear: employees in Japan, as in many other countries, often struggle with a culture of long working hours and the expectation of remaining connected even during personal time.

Microsoft aimed to test whether reducing workdays—and consequently reducing digitally connected work time—would harm productivity or, in fact, improve it.

The Action Taken: The company gave all 2,300 employees Fridays off throughout the month, amounting to five consecutive paid Fridays, with no reduction in salary. Management also encouraged teams to shorten and streamline meetings, for example by setting a maximum meeting length of 30 minutes, allowing the four working days to be used more effectively. In addition, Microsoft offered financial support for family travel during the extended weekends, providing up to the equivalent of USD 920 for employees who chose to travel with their families. This incentive was designed to encourage employees to use the extra time for genuine rest and personal wellness.

Results: The outcomes were striking and attracted global attention. Productivity increased by 40 percent during that month compared with previous periods, a result that surprised even company leadership. Condensing five days of work into four encouraged employees to focus more sharply and work more efficiently. Shorter meetings and clearer communication helped teams accomplish more in less time.

Employee satisfaction rose sharply. Surveys showed that 92 percent of employees enjoyed the shorter workweek and wanted it to continue, reflecting a strong increase in morale and job satisfaction. The company also recorded a 25 percent decrease in absenteeism and sick leave requests during the experiment. Additional rest appeared to improve overall wellness and reduce stress-related fatigue.

Environmental and cost-related benefits were also recorded. Electricity use in company offices dropped by approximately 23 percent due to an extra day of closure, and paper printing declined by around 59 percent, indicating a shift toward more efficient and sustainable work practices.

Takuya Hirano, President of Microsoft Japan, summarized the experience with a simple message: "Work a shorter time, rest well, and learn a lot." His words reflected the idea that allowing employees more time for life enhances energy, motivation, and long-term learning.

Overall, the experiment provided strong evidence that supporting digital wellness, by reducing excessive digital demands and constant connectivity, benefits both employees and organizations. Employees experienced better mental and physical health, while the organization benefited from higher performance and productivity. The results also inspired other companies worldwide, including Perpetual Guardian in New Zealand, which adopted a four-day workweek and reported similar improvements in work–life balance and reduced stress among employees.

The Right to Disconnect: A "No Emails After Hours" Culture

Another inspiring example comes from European initiatives designed to protect employees' digital downtime. In Germany, Volkswagen took an early step as far back as 2011, responding to employee complaints about the pressure of after-hours email communication.

In cooperation with the labor union, the company introduced a policy that stops the email server from delivering messages to employees' phones 30 minutes after the workday ends and resumes delivery only 30 minutes before the next workday begins. This measure ensured that employees would not receive work emails overnight, except in urgent or exceptional cases.

The goal was to preserve employees' personal time and allow them to mentally disconnect from work after long hours on the job.

The outcome was overwhelmingly positive. Employees welcomed the policy and felt that the company respected their private lives. Many reported that the anxiety of checking emails at night disappeared entirely, allowing them to spend quality time with family or personal interests without stress. Management also observed that employees returned to work feeling more rested, focused, and engaged. The arrangement created a sense of fairness: work during work hours, and rest without interruption afterward.

Today, the concept of a "right to disconnect" has become increasingly widespread across Europe. In France, a law introduced in 2017 formally granted employees the right to ignore work emails and messages outside official working hours without penalty. Companies in countries such as Belgium, Italy, and Spain have since adopted similar internal policies. Together, these developments reflect growing recognition that digital exhaustion is a serious workplace issue and that sustained productivity depends on a balanced and protected workforce.

A Work Environment That Supports Digital Wellness: Diverse Examples

Beyond formal policies, some organizations cultivate digital wellness through everyday workplace practices. A few illustrative examples include:

A Startup in Sweden: The company designates a technology-free lunch hour each day. During this time, laptops and phones are not permitted, and employees are encouraged to eat together, talk, or rest. Management describes this break as a mental "recharge" that improves creativity and productivity in the second half of the day.

A government entity in the UAE: It launched an internal campaign titled "Connect Without Technology on a Set Day," selecting one day each month when meetings are held without electronic devices. Employees discuss ideas using only paper and pens, fostering direct interaction. Many participants reported livelier discussions and renewed creativity, reminding them that meaningful ideas often emerge away from screens.

A global consulting firm: The firm introduced a voluntary "digital detox" program as part of its wellness strategy. Employees who enroll receive one-on-one coaching from specialists in time management and healthy technology use. Participants create personalized plans to reduce phone dependency or establish screen-free routines before bedtime. Follow-up surveys showed that participants reported stress reductions exceeding 25 percent, along with higher job satisfaction.

These examples demonstrate that a workplace culture centered on digital wellness creates a win–win outcome. Employees feel healthier, calmer, and more valued, while organizations benefit from increased loyalty, creativity, and productivity. Companies that recognize employees as human beings with real needs—rather than machines expected to remain connected at all times—are more likely to earn long-term commitment. The evidence is clear: as the Microsoft Japan case showed, giving people more rest can lead to better results, not lower performance. Sometimes, reducing digital work is precisely what improves the quality of work itself.

From Inspiring Models to a Comprehensive Culture

In this chapter, we explored a wide range of real-world case studies from across the globe, all conveying a single message: digital wellness is not an abstract ideal but a practical and achievable goal. We saw how individuals regained control over their lives by confronting digital overuse with determination. We observed families restoring closeness and balance through simple rules and shared responsibility. We examined schools and universities that nurtured awareness in younger generations, producing students who are more mindful of how technology shapes their daily lives.

Finally, we saw how institutions and companies that actively protected employees' digital wellness earned greater loyalty and higher productivity, proving that investing in psychological and digital health strengthens organizational performance. Together, these examples show that when digital wellness becomes a shared cultural value, it can transform lives, families, institutions, and entire communities.

These real-world examples, from the UAE, Saudi Arabia, and Egypt to Canada, Sweden, Japan, and beyond, send a powerful message of hope and motivation. They confirm that digital challenges that may seem impossible to solve can be addressed through gradual, intentional steps when conviction and determination are present. More importantly, they reveal that meaningful change never happens in isolation. Every success story emerged from collective effort and shared responsibility within a surrounding community. Violet would not have succeeded without her school's initiative and her family's support. The family that overcame device chaos found strength through a national awareness campaign and a supportive social environment. The school that reshaped students' digital habits succeeded because parents cooperated and students themselves understood the value of change. The company that granted employees the right to rest and be treated humanely inspired other organizations to follow its example.

We can imagine this growing influence as a positive snowball effect. One success story inspires ten more, and every effective initiative in a school or workplace encourages other institutions to try the same approach. In this way, a culture of digital wellness gradually expands until it becomes part of a society's shared awareness. What once felt unusual, choosing to turn off a phone in the evening, limiting technology during certain school activities, or restricting after-hours work communication, slowly becomes normal, accepted, and even expected.

In an era where technology reaches into every aspect of life, this culture is needed more than ever. Digital wellness does not mean rejecting technology or resisting progress. It means using technology consciously and in moderation, benefiting from its advantages for ourselves and our children without sacrificing health, balance, or happiness. The stories shared in this chapter demonstrate that the true solution lies in balance: no excess that erodes real life, and no neglect that disconnects us from modern opportunities.

As we conclude, it is worth reflecting on the words of a digital education expert involved in the "Digital Bridge" project: "Change happens when people feel that someone sees them and understands their struggle, and when they are given support, a healthy structure, and a shared purpose." This insight captures exactly what we witnessed in these real-world examples. When people felt understood rather than judged, they became more willing to seek support and embrace change. Small communities, a family, a classroom, a work team, set shared goals and moved toward them step by step.

Let us also draw inspiration from these models. Each of us can begin with a simple step at home or at work: a family conversation about digital habits, a proposal for a day without electronic meetings, or even a personal "Screen-Free Saturday." Who knows, your small action might become the spark that creates a new success story, inspiring others along the way. A culture of digital wellness grows through individuals, families, and institutions, one after another. And since we have already witnessed its positive impact in these real examples, there is no doubt that the future will bring many more stories like them—and perhaps your story will be among the next inspiring ones.

Chapter Ten:
"Laws and Regulations Related to Digital Wellness"

Why is the Law Considered Part of Digital Wellness?

The law is not separate from our daily lives. It is the protective framework that keeps our digital lives within healthy boundaries. It relates to the data we collect, or the data collected by us, or how apps and platforms use that data, and when we need to make decisions, such as: Can this app track my child without permission? Can a company send intrusive ads? Do I have the right to request the deletion of my information?

All of these questions are connected to rights protected by law in a world where nearly everyone is constantly connected. Neglecting control over these matters has a direct impact on mental health, such as cognitive fatigue caused by excessive advertising, social anxiety resulting from loss of digital privacy, and mental distraction due to continuous pressure.

Modern laws such as the GDPR in the European Union, or PDPL regulations in the Gulf, regulate the relationship between people and technology, whether at the individual level, in educational settings, at work, or within government services. At their core, these laws function as tools for digital wellness because they link the use of technology to principles that must be upheld, including transparency, consent, limits, and the right to disconnect.

Therefore, limiting email after working hours is not merely a lifestyle preference. It has become part of a legal and organizational approach known as the "right to disconnect," which France and Germany have adopted, and which many major companies have incorporated as a foundation for protecting employees' mental wellness.

In short, the law gives us the right to say no and restores our ability to control our digital environment. It ensures that apps do not control us, but that we control them.

Core Principles in Modern Digital Regulations

In brief, the shared principles of digital protection across most contemporary laws can be summarized in the following points:

Principle	Explanation
Data Minimisation (Minimum Necessary Data)	Collect, use, or store only the minimum amount of data that is strictly necessary.
Informed Consent	User consent must be clear, explicit, and obtained in advance, with the right to withdraw it at any time.
Data Subject Rights	Includes the right to access, rectify, erase (the right to be forgotten),

	object, and request data portability.
Enhanced Protection for Children	Restricts the collection, analysis, or marketing use of children's data, with additional approval and safeguards.
Secure and Cross Border Data Transfers	Any transfer of data outside national borders is subject to controls, such as adequacy decisions, restrictions, and standard contractual clauses.
Breach Notification	Organizations must notify the regulator and the affected individual(s) of a data breach within a specified timeframe.
Deterrent Penalties	Imposes significant financial fines on organizations or apps that violate rights or breach legal obligations.

These principles have become the benchmark for any country seeking to protect personal data and a central pillar in every discussion of digital wellness. They strengthen user control while reducing the psychological strain caused by constant notifications, tracking, and targeted advertising. In practical terms, they help prevent technology from becoming a permanent source of distraction and negative emotional pressure.

Legal Decisions and Regulatory Frameworks in the Gulf States

This section reviews key laws in Gulf countries that relate to digital wellness as an extension of data security, privacy protection, and user rights.

United Arab Emirates (PDPL, Federal Decree Law No. 45 of 2021)

The law was issued in December 2021. Most provisions entered into force in January 2022, with practical implementation beginning in early 2023.

It grants data subjects comprehensive rights, including access, correction, erasure, data portability, and the right to object to unnecessary processing. Data subjects also have a clear legal entitlement to request enforcement free of charge within a reasonable period, typically between 30 and 45 days.

Data minimization is mandated, and tracking a child or targeting them with advertising is considered a violation unless valid, informed, and transparent consent is obtained.

The main legislative text (FDL 45 of 2021) awaited its executive regulations until early 2025, meaning that practical procedures may still be adjusted or further clarified. Organizations are therefore advised to prepare early in order to keep pace with upcoming developments.

Saudi Arabia (PDPL, published September 2023)

The law officially entered into force on 14 September 2023, and organizations were granted a one-year compliance grace period until 14 September 2024 for full implementation.

The Saudi Data and AI Authority (SDAIA) issued the executive regulations on 7 September 2023, introducing controls on cross-border data transfers, consent mechanisms, and mandatory Privacy Impact Assessments. These assessments require organizations to evaluate privacy risks and potential impacts before processing data.

As of 14 September 2024, enforcement began in practice, including the application of judicial fines and administrative measures against non-compliant entities.

Oman (Royal Decree No. 6 of 2022)

Issued in February 2022, the law became effectively applicable on 13 February 2023. The executive regulations were published on 29 January 2024, and institutions were granted until February 2025 to achieve full compliance.

The regulations require entities to respond to data subject requests, including access, deletion, and data portability, within 45 days. Individuals may submit a complaint to the competent authorities if there is no response within 60 days, providing users with legally binding protection tools.

Qatar (Law No. 13 of 2016 on Personal Data Privacy Protection)

Qatar's law is widely regarded as the first comprehensive personal data protection framework in the Gulf. It was issued in November 2016 and entered into force in 2017.

The law requires explicit consent, particularly parental consent when processing children's data, includes penalties for violations, and is overseen by the Ministry of Communications and Information Technology.

In recent years, Qatar has moved toward stronger enforcement, including actions against entities that fail to comply with regulations governing sensitive data or engage in indiscriminate marketing aimed at children.

Bahrain (Law No. 30 of 2018)

Issued on 19 July 2018 and enforced from August 2019, Bahrain also established the Personal Data Protection Authority (PDPA) to supervise implementation. This institutional structure has made the framework more comprehensive and enforceable than several earlier regional approaches.

The law closely mirrors the European GDPR, particularly in its emphasis on explicit consent, individual control over personal data, defined user rights, restrictions on cross-border transfers, and guidance for handling sensitive information.

Kuwait

Kuwait has not yet adopted a single comprehensive data protection law. However, several scattered provisions address privacy and data issues, including the Electronic Transactions Law (2014), the Communications Authority decision (2020) on data protection in telecom services, and the Cybercrime Law (2015).

As a result, it is generally recommended that institutions—especially schools and companies—follow international best practices aligned with PDPL and GDPR standards, even in the absence of a unified enforcement framework in Kuwait.

International Policies and Laws Influencing Digital Wellness

European Union: Digital Services Act (DSA)

The DSA has been effectively applicable since 17 February 2024 to digital platforms and applications that provide services to EU citizens. It obliges platforms to protect users, particularly children, from harmful content and addictive behaviors, and to provide accessible tools for complaints and reporting harmful or misleading material. These measures represent a concrete step toward improving digital wellness.

On 14 July 2025, the European Commission issued a final guide on protecting minors under Article 28 of the DSA. The guide emphasized the importance of "default privacy settings for minors," called for banning advertising based on tracking mental activity, and urged restrictions on addictive design features in platforms accessible to children.

Pilot projects were launched in countries such as France, Spain, Italy, and the Netherlands to verify users' ages before granting access to inappropriate content. These initiatives form part of broader efforts to protect younger users, with some countries testing age-verification apps as an interim solution ahead of the full launch of the European Digital Identity Wallet in 2026.

France: Ban on Phones in Schools

France adopted Law No. 2018-698 on 3 August 2018, banning the use of mobile phones and smart devices in primary and lower secondary schools, up to the age of 15, throughout the school day, including breaks and school trips.

Between 2024 and 2025, France announced an expansion of the ban across all lower secondary schools, requiring students to place phones in locked boxes at the start of the day. The stated objectives include restoring concentration, encouraging direct social interaction, and reducing cyberbullying within schools.

Europe: The "Right to Disconnect" and More Humane Workplaces

In Germany, Volkswagen was among the first companies to limit non-urgent work emails after working hours, beginning in 2012, as part of broader employee wellness and digital stress reduction measures.

This practice aligns with the wider concept of the "right to disconnect," later reflected in European policy frameworks, most notably France's 2017 legislation. This right protects employees' digital downtime, supporting mental health and long-term productivity.

Recent Trends (2024–2025), Shaping Digital Balance

AI regulation (EU AI Act): Implementation began in 2024, addressing high-risk design practices such as addictive recommendation systems and interfaces that exploit cognitive dependency. These measures encourage more responsible AI design and reduce emotionally manipulative engagement. While not yet fully adopted in the Gulf, similar regional directions are emerging.

Shorter Workweek Experiments (Four-Day Week): Microsoft Japan's August 2019 trial reported a 40 percent productivity increase, reduced electricity consumption, and higher employee satisfaction. European companies have continued testing similar models, including firms such as Awin and Civo in the UK, reporting improved wellness and fewer absence days without loss of productivity.

European Commission Initiatives to Protect Children (+BIK): These initiatives focus on improving children's digital environments by promoting age-appropriate content, default privacy settings, and children's rights to access their data and communicate with parents or guardians.

International Education Initiatives: Examples include the University of Michigan's peer-to-peer digital wellness program, which connects university students with middle school students to promote healthy technology habits. Another example is Sweden's school-day mobile phone ban, introduced to enhance concentration and overall student wellness.

How can you Strengthen your Digital Wellness using Legal Tools?

For Families and Individuals

Read an application's privacy policy before allowing children to download it, and ensure there is a clear option to disable activity tracking and limit targeted advertising. These rights are legally supported under the UAE, Saudi, and Bahraini PDPL frameworks, as well as the European GDPR.

Set household digital rules aligned with national guidance, such as banning phone use at the dining table or enabling Night Mode and Do Not Disturb settings after 9:00 pm. These practices are reinforced by child-protection frameworks that promote a safer and more balanced digital environment.

Use legal rights when necessary, such as requesting deletion of a child's data from platforms no longer in use or reporting intrusive advertisements aimed at children. Most data protection laws across the Gulf, Jordan, and Europe provide clear legal grounds for such actions.

For Institutions and Schools

Develop a clear and simplified privacy policy in both Arabic and English that explicitly defines what data is collected, the purposes for its use, the names of service providers, and the rights of data subjects.

Make parental consent mandatory when enrolling children in applications, in strict accordance with Qatar's data protection law, European child protection frameworks, and the GDPR. Maintain documented copies of consent forms and renew or update them on an annual basis.

Avoid adopting technologies that analyze the relationship between user clicks and psychological states, or AI systems designed to encourage excessive or unnecessary engagement. Such designs are either prohibited or subject to heightened regulatory scrutiny under frameworks such as the Digital Services Act and the EU AI Act.

Workplace and Office Environment

Adopt a clear and enforceable policy governing email and meeting practices. Define firm cut-off hours at the end of the workday, for example treating "no emails after 7:00 pm" as a standard rule, and apply it consistently during weekends and public holidays.

Promote a workplace culture in which disabling notifications after work hours is recognized as a legitimate right rather than an exception. Even where this is not yet legally binding, it strongly supports mental disconnection, recovery, and sustainable performance.

Provide regular training on responsible application use, guiding employees on how to adjust privacy settings on their devices and maintain digital balance. This approach also aligns with PDPL compliance expectations, which require organizations to raise awareness among staff about data protection and responsible technology use.

How Laws Affect the Quality of Our Digital Lives

By safeguarding user rights, laws reduce notification overload and help alleviate the stress caused by constant digital connectivity.

When phones are restricted in classrooms or within the home, legal provisions and civic commitments reinforce concentration, strengthen real-life relationships, improve sleep quality, and support healthy personal development.

Laws also act as drivers of broader societal change. Developers and technology providers are compelled to redesign platforms to comply with privacy requirements, making responsible system design more effective than relying on individual self-control alone.

When these laws are translated into everyday practices in schools and workplaces, family and organizational wellness improves. People feel that their time is protected and respected as a principle, rather than treated as a resource for continuous monitoring and extraction.

Law as a Catalyst for Change, Not an Obstacle

When laws are crafted not merely for oversight or surveillance, but with a clear vision of protecting individuals within a digital society, they become an invisible constitution for digital wellness. Across Europe, legal frameworks have been transformed into daily practices, including age verification before accessing harmful content, banning phones in schools, and granting employees the right to disconnect after working hours.

In the Gulf region, momentum toward full implementation is steadily accelerating. The UAE, Saudi Arabia, Bahrain, and Oman have begun enforcing their laws, even if some remain in transitional stages. What matters most is recognizing that these laws are not the end of the journey, but its beginning. Digital wellness is a shared responsibility among individuals, families, schools, companies, and digital platforms, shaping how societies live with healthier technological balance.

From this perspective, a digital wellness approach extends beyond managing screen time alone. It also involves regulating data overload and redefining our relationship with technology through clear conditions: legal conditions, psychological conditions, and social conditions.

Start today with a small step that aligns with the law. Read an application's privacy policy, strengthen privacy settings for your child, or propose a "phone-free day" at school. The law gives you the right to act, but real transformation begins with individual and collective choice.

References

1. Zablotsky, B. (2024). Daily Screen Time Among Teenagers: United States, July 2021-December 2023. National Centre for Health Statistics.

2. Bakhai, A. (2022). Work-related communications overwhelm staff: 82% support need for guidance to protect emotional wellbring. PMC.

3. Karanika-Murray, M. et al. (2025). Digital Technologies for Well-being at Work. Horizon Scanning Study.

4. Pielot, M. & Rello, L. (2016). Productive, Anxious, Lonely 24 Hours Without Push Notifications.

5. Dai, Y. & Ouyang, N. (2025). Excessive Screen Time is Associated with Mental Health Problems and ADHD in US Children and Adolescents [Preprint].

6. Soffer, I. et al. (2025). The Anxious Generation: How the Great Rewiring of Childhood Is Causing an Epidemic of Mental Illness.

7. Pissarides Review into the Future of Work and Wellbeing, Institute for the Future of Work (2024). Does technology use impact UK workers' quality of life?

8. CIPD & Simply Health (2023). Health and Well-being at Work.

9.   Surgeon General of the USA (2023). Social Media and Youth Mental Health.

10.   Financial Times (Oct 2024). Teenage social media use strongly linked to anxiety and depression.

11.   CDC/NCHS via CBS News (Oct 2024). Half of teens with 4+ hours screen time report anxiety or depression.

12.   pewresearch.org

13.   livenowfox.com

14.   themuse.com

15.   unesco.org

16.   datareportal.com

17.   samenacouncil.org

18.   worldbank.org

19.   businessinsider.com

20.   www.nippon.com/ar

21.   www.thenationalnews.com/news/uae/2025/01/22/chil

22.   d-safety-online

23.   saudipedia.com

24.   National Center for Health Statistics (United States) (2024).

25. NHS (United Kingdom) (2022). Digital communication after hours.

26. University of Oxford (2024). The relationship between adolescents' use of social media and levels of anxiety and depression.

27. www.bshra.com

News Sources

28. **CBS News** (2024, October 31)

29. **The Guardian** (2024, March 12).

www.ingramcontent.com/pod-product-compliance
Lightning Source LLC
Chambersburg PA
CBHW040906110726
48005CB00006B/814